W0111873

Fundamentals of Natural Gas Processing

Fundamentals of Natural Gas Processing

Editors

Pankaj Kurulkar and Deepa Honap

scitus
academics

Fundamentals of Natural Gas Processing

Edited by **Pankaj Kurulkar and Deepa Honap**

Printed in 2017

ISBN: 978-1-68117-378-8

Library of Congress Control Number: 2015936531

© 2016 by
SCITUS Academics LLC,
616, Corporate Way, Suite 2, 4766,
Valley Cottage, NY 10989

www.scitusacademics.com

This book contains information obtained from highly regarded resources. Copyright for individual articles remains with the authors as indicated. All chapters are distributed under the terms of the Creative Commons Attribution License, which permits unrestricted use, distribution, and reproduction in any medium, provided the original author and source are credited.

Notice

Reasonable efforts have been made to publish reliable data and views articulated in the chapters are those of the individual contributors, and not necessarily those of the editors or publishers. Editors or publishers are not responsible for the accuracy of the information in the published chapters or consequences of their use. The publisher believes no responsibility for any damage or grievance to the persons or property arising out of the use of any materials, instructions, methods or thoughts in the book. The editors and the publisher have attempted to trace the copyright holders of all material reproduced in this publication and apologize to copyright holders if permission has not been obtained. If any copyright holder has not been acknowledged, please write to us so we may rectify.

Contents

Preface

Natural-gas processing is a complex industrial process designed to clean raw natural gas by separating impurities and various non-methane hydrocarbons and fluids to produce what is known as pipeline quality dry natural gas. Natural-gas processing begins at the well head. The composition of the raw natural gas extracted from producing wells depends on the type, depth, and location of the underground deposit and the geology of the area. Oil and natural gas are often found together in the same reservoir. The natural gas produced from oil wells is generally classified as associated-dissolved, meaning that the natural gas is associated with or dissolved in crude oil. Natural gas production absent any association with crude oil is classified as "non-associated."

Editor

Latest Development on Membrane Fabrication for Natural Gas Purification: A Review

Dzeti Farhah Mohshim, Hilmi bin Mukhtar,
Zakaria Man, and Rizwan Nasir

Chemical Engineering Department, Universiti Teknologi Petronas,
Bandar Seri Iskandar, Perak Darul Ridzuan, 31750 Tronoh, Malaysia

ABSTRACT

In the last few decades, membrane technology has been a great attention for gas separation technology especially for natural gas sweetening. The intrinsic character of membranes makes them fit for process escalation, and this versatility could be the significant factor to induce membrane technology in most gas separation areas. Membranes were synthesized with various materials which depended on the applications. The fabrication of polymeric membrane was one of the fastest growing fields of membrane technology. However, polymeric membranes could not meet the separation performances required especially in high operating pressure due to deficiencies problem. The chemistry and structure of support materials like inorganic membranes were also one of the focus areas when inorganic membranes showed some positive results towards gas separation. However, the materials are somewhat lacking to meet the separation performance requirement. Mixed matrix membrane (MMM) which is comprising polymeric and inorganic membranes presents an interesting approach for enhancing the separation performance. Nevertheless, MMM is yet to be commercialized as the material combinations are still in the research stage. This paper highlights the potential promising areas of research in gas separation by taking into account the material selections and the addition of a third component for conventional MMM.

INTRODUCTION

Natural gas can be considered as the largest fuel source required after the oil and coal [1]. Nowadays, the consumption of natural gas is not only limited to the industry, but natural gas is also extensively consumed by the power generation and transportation sector [2]. These phenomena supported the idea of going towards sustainability and green technology as the natural gas is claimed to generate less-toxic gases like carbon dioxide (CO_2) and nitrogen oxides (NO_x) upon combustion as shown in Table 1 [3].

Table 1: Fossil fuel emission levels (pounds per billion Btu of energy input)

Fuel sources/pollutant (pound/BTU)	Natural gas	Oil	Coal
Carbon dioxide	117,000	164,000	208,000
Carbon monoxide	40	33	208
Nitrogen oxides	92	448	457
Sulphur dioxide	1	1,122	2,591
Particulates	7	84	2,744
Mercury	0.000	0.007	0.016

However, pure natural gas from the wellhead cannot directly be used as it contains undesirable impurities such as carbon dioxide (CO_2) and hydrogen sulphide (H_2S) [4]. All of these unwanted substances must be removed as these toxic gases could corrode the pipeline since CO_2 is highly acidic in the presence of water. Furthermore, the existence of CO_2 may waste the pipeline capacity and reduce the energy content of natural gas which eventually lowers the calorific value of natural gas [5].

Conventionally, natural gas treatment was predominated with some methods such as absorption, adsorption, and cryogenic distillation. But these methods require high treatment cost due to regeneration process, large equipments, and broad area for the big equipments [6]. With the advantages of lower capital cost, easy operation process, and high CO_2 removal percentage, membrane technology offers the best treatment for natural gas [6]. Natural gas is expected to contain less than 2 vol% or less than 2 ppm of CO_2 after the natural gas treatment in order to meet the pipeline and commercial specification [7]. This specification is made to secure the lifetime of the pipeline and to avoid an excessive budget for pipeline replacement.

Membrane technology has received significant attention from various sectors especially industries and academics in their research as it gives the most relevant impact in reducing the environmental problem and costs. Membrane is defined as a thin layer, which separates two phases and restricts transport of various chemicals in a selective manner [8]. Membrane restricts the penetration of some molecules that have bigger kinetic diameter. The commercial value of membrane is determined by the membrane's transport properties which are permeability and selectivity. Major gap of the existing technologies is limited to low CO_2 loading (<15 mol%). Ideally, we required high permeability and high selectivity of membrane, but, however, most membranes exhibit high selectivity in low permeability and vice versa which make this is as a major tradeoff of membranes, and none of these technologies are yet to treat natural gas containing high CO_2 (>80 mol%) [9].

MEMBRANE TECHNOLOGY DEVELOPMENT

Early Membrane Development

Membrane technology has been started as early as in 1850 when Graham introduced the Graham's Law of Diffusion. Then, gas separation utilization in membrane technology has been commercialized in late 1900's. Permea PRISM membrane was the first commercialized gas separation membrane produced in 1980 [2]. Summary of early development of membranes is shown in Figure 1. This innovation has led to the further membrane gas separation development. A lot of studies done by the researchers for various gas separation mostly focus on the natural gas purification.

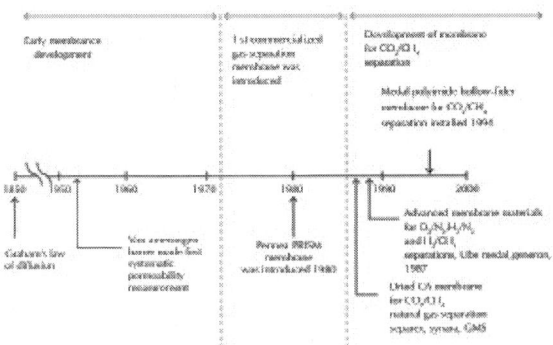

Figure 1: Membrane development timeline.

Development of membrane for CO_2/CH_4 separation has been started since early 1990's. Numbers of membranes were fabricated using different kind of materials in the early stage of this membrane gas separation. The desirable material selected must be well suited to the separation performance by which mean separation of gases works contrarily in different materials. Excellent gas membranes separation should have the characteristic of high separation performance with reasonable high permeability, high robustness, chemically, thermally, and mechanically good and rational production cost [10, 11]. Two types of materials are practically used in gas separation: polymeric membrane and inorganic membrane and the comparison of both polymeric and inorganic membranes is showed in Table 2.

Table 2: Comparison between polymeric and inorganic membranes

	Polymeric membranes	Inorganic membranes
Materials	Present in either rubbery or glassy type which depends on the operating temperature [12].	Made from inorganic-based material like glass, aluminium, and metal [13].
Characteristics	(i) Polymer is more rigid and hard in glassy state while in rubbery state it is more soft and flexible. (ii) Glassy polymeric membranes exhibit higher glass transition temperature compared to rubbery membranes, and glassy types tend to have higher CO_2/CH_4 selectivity [14].	(i) Able to withstand with solvent and other chemicals and also susceptible to microbial attack. (ii) Comprise significantly higher permeability and selectivity, but they are also more resistant towards higher pressure and temperature, aggressive feeds, and fouling effects [15].
Disadvantages	(i) May have plasticization problem when handling high CO_2. (ii) Presence of CO_2 may result in membrane performance reduction at certain elevated pressure. (iii) As the membranes expose to CO_2, polymer network in the membrane will swell, and segmental mobility will also increase which consequently cause a rise in permeability for all gas components [16]. (iv) The components with low permeability characteristic will experience more permeability increment; thus, the selectivity of the membrane will definitely decrease [17–19].	(i) Inherent brittleness characteristic. (ii) Performed well under low pressure which does not suit the natural gas well which required high pressure for the exploration. (iii) High production cost which seems not practical for large industrial applications [20].

Examples	Polyethylene (PE), poly(dimethylsiloxane) (PDMS), polysulfone (PSU), polyethersulfone (PES), polyimide (PI) [21], polycarbonate [22], polyimide [23], polyethers [24], polypyrrolones [25, 26], polysulfones [27], and polyethersulfones [28].	Aminoslicate membrane [29], carbon-silicalite composite membrane [30], MFI membranes [31], and microporous silica membranes [32].

Gas separation using polymeric membranes has taken its first commercial scale in late 1970's after the demonstration of rubbery membranes back in 1830's [33]. Literally, the permeability of gas in a specific gas mixture varies inversely with its separation factor. The tighter of molecular spacing it has, the higher the separation characteristic of the polymer, but, however, as the operating pressure increases, the permeability is decreasing due to experiencing lower diffusion coefficients [34]. Polymeric membranes that are commercially available for CO_2/CH_4 separation include polysulfone (PSU), polyetehrsulfone (PES), polyamide (PI) and many more. Generally, as the permeability of the gas increases, the permselectivity was attended to decrease in most cases of polymeric membranes [23].

Inorganic membrane like SAPO-34 could give higher separation performance compared to the polymeric membrane, but the separation performance is inversely proportional to the pressure loaded. This observation may create problem when we deal with high pressure natural gas well. The performance of both organic and inorganic membrane is summarized in Robeson's plot as in Figure 2 [35].

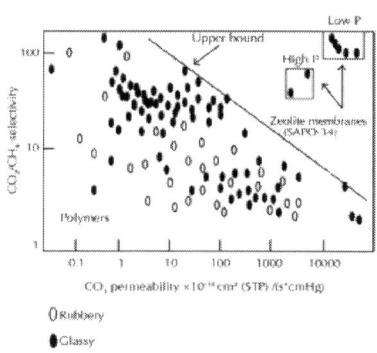

Figure 2: Zeolite (SAPO-34) membrane performance in Robeson's plot.

Conventional Mixed Matrix Membrane

A lot of researches have been done to satisfy the needs of gas separation requirement through both polymeric and inorganic membranes. The deficiencies of these membranes have driven the researchers to develop an alternative material for membrane which is more mechanically stable and economic viable, and most important is having high separation performance. The combination of organic and inorganic material which is known as mixed matrix membrane (MMM) was then proposed in idea to get a better membrane gas separation performance at reasonable price [36]. The fabrication of MMM was a promising technology as this composite material has improved its mechanical and electrical properties [37], and it combines the exceptional separation ability and pleasant stability of molecular sieves with better processability of organic membrane [38]. The MMM is characterized by dispersing the inorganic material into the continuous phase of polymeric material which can be almost any polymeric material such as polysulfone, polyimide, and polyethersulfones [39, 40].

Various membrane materials can be selected based on the process requirement. Selected materials can be "tailored-made" in order to meet the specific separation purpose in a wide range of application [39]. There were many attempts of developing polymer-inorganic membrane that started few decades back then.

Based on Table 3, this was observed that the selection of materials is important, and it depends on the system requirement. Higher intrinsic diffusion selectivity characteristic of glassy polymer makes this material better than rubbery polymer [56]. Although MMM has proven an enhancement of selectivity, it was noticed that most MMMs were endured with poor adhesion between the organic matrix and inorganic particles [55]. Even MMM fabrication does have its disadvantages, but the research of MMM with different materials is worth to work on since it has proven its ability to have high separation performance.

Table 3: Few researches of mixed matrix membranes

Year	Mixed matrix membrane (MMM)		Observations	Ref.
	Organic	Inorganic		
1973	Silicon rubber	Molecular sieves	Poor adhesion of organic and inorganic selected leads to poor separation performance. This poor interaction of both materials may result in nonselective voids present at the interface which consequently causes insufficient membrane performance [41–43].	[44]
1992	Polydimethylsiloxane (PDMS)	Silicalite-1, 13X, KY, and zeolite-5A	Zeolite like silicalite-1, 13X, and KY have enhanced the separation performance of poorly selective rubbery membrane for the carbon dioxide (CO_2) and methane (CH_4) mixture.	[45]
	Propylene diene rubber (EPDM)		Zeolite-5A showed no change in gas selectivity with decrease permeability due to impermeable characteristic towards CO_2.	
2000	Cellulose acetate (CA)	Silicalite, NaX, and AgX	Silicalite did in fact reverse the selectivity of CA membrane from H_2 to CO_2 for CO_2/H_2 separation.	[46]
2000	Polyvinyl acetate	4A	Formation of chemical bonds gave good adhesion, but there is still nonselective "leakage" from the existence of nanometric region.	[47]

2003	Matrimid	Carbon molecular sieves	Selectivity of CO_2/CH_4 mixture has increased up to 45%. Zeolites loading also affects both gas permeability and gas mixture selectivity. There were also a number of records where permeability increased with selectivity decreased as the zeolites loading was increased [48, 49] and vice versa [42].	[50]
2006	Polyethersulfone (PES)	Zeolite 4A	Due to low mobility of the polymer chain in glassy polymer such as to prevent them to completely cover the zeolites surface which resulted in void interface [51, 52].	[53]
2001	Polyimide (PI)	Zeolite 13X		[54]
2008	Polycarbonate	Zeolite 4A		[55]

Recent Development of Membrane Gas Separation

Ionic Liquid-Supported Membrane (ILSM)

In recent years, many researches have been evaluated on the ionic liquid supported membrane (ILSM) for gas separation membrane since ionic liquids are known materials that could dissolve CO_2 and stable at high temperature ranges [57]. To be specific, ionic liquids are molten salt that are liquid at room temperature [58]. Furthermore, ionic liquids are of particular interest for membrane gas separation application as they are inflammable, negligible vapour pressure, and nonvolatile which make them also known as "green" solvents [58–60]. Extensive researches have been carried out to develop room temperature ionic liquid (RTIL)-based solvents for CO_2 separation with various types of ionic liquids such as pyridinium and imidazolium based. Among RTILs tested, imidazolium-based RTIL was chosen as the most feasible solvent for CO_2 separation as they are commercially viable and easily tunable by tailoring the cation and anion to meet the system requirements [60].

ILSMs have been proven that they offered an increase in permeability that outperforms many neat polymer membranes. ILSMs synthesized from poly(vinylidene fluoride) (PVDF) and 1-butyl-3-methylimidazolium tetrafluororate (BMImBF$_4$) showed high permeation performance of CO_2 and mechanically stable while operating at high pressure condition [63]. The consumption of RTILs showed an increment especially for 1-R-3-methylimidazolium (R-mim)-based RTILs as this type is preferable due to its properties of less viscous compared to other RTILs. In addition, gases like CO_2, nitrogen (N$_2$), and other hydrocarbons demonstrated high solubility in Rmim-based RTILs [64, 65]. Besides, the use of Rmim-based RTILs could calculate the latent permeability and selectivity of the mixture of given gases by using the molar volume of these RTILs [60]. RTIL can be functionalized and set up in according to the system requirement and application, and these researches could be good benchmark for designing the functionalized RTIL efficiently as showed in Table 4.

Table 4: Effects of ionic liquid functionalization

Functionalization	Effects
Nitrile and alkyne group	(i) Gas solubility and separation performance have been tailored.
	(ii) Functionalized RTIL solvents displayed a decreasing in CO_2, N$_2$, and CH$_4$ solubility, but, however, the selectivity of CO_2/N$_2$ and CO_2/CH$_4$ increased when compared to the nonfunctionalized RTIL [61].
Temperature	(i) As the temperature increases, the CO_2 solubility is decreasing while the CH$_4$ solubility remains unchanged.
	(ii) The ideal solubility selectivity of mix gases for CO_2/N$_2$, CO_2/CH$_4$, and CO_2/H$_2$ increased as the temperature decreased [62].

Polymerized Room Temperature Ionic Liquid Membrane (Poly RTIL))

Comparatively, RTIL especially imidazolium based can be also polymerized into a solid, dense, and thin film membrane due to their modular nature [66–68]. It was a successful breakthrough when the researcher found that polymer from ionic liquid monomer had higher CO_2 absorption capacity with faster absorption and desorption rate compared to the neat RTIL [69]. Moreover, poly(RTIL) is also attributed with higher mechanical strength [66]. These characters have proven that polymerized ionic liquid (poly(RTIL)) is also a promising material for membrane gas separation. Polymerization of RTIL monomer by varying the n-alkyl length also showed a pleasant result when increase of permeability of given gases like CO_2, N_2, and methane (CH_4) was observed as the n-alkyl group was lengthened [68]. Additionally, poly(RTIL) is also up to extend when it practically absorb about twice as much CO_2 as their liquid analogue which makes it much better than molten RTIL [68]. Apparently, performance of poly(RTIL) also depends on the substituent attached to it. In a research done on the inclusion of a polar oligo(ethylene glycol) on the cation side of imidazolium-based RTIL, the separation selectivity has seemed to increase [70].

As discussed earlier, mixed matrix membrane is a known membrane that composed of a compatible organic-inorganic pair which demonstrated having good separation properties subject to no interfacial adhesion problem. The improvement of separation performance is expected in an MMM comprising poly(RTIL) (polymer matrix) and zeolite (inorganic). In a very recent work, the benefit of MMM has become an idea to the researcher in ionic liquid membrane field. Hudiono and his coworkers have introduced a three-component mixed matrix membrane by utilizing the poly(RTIL), RTIL, and zeolite [71]. Their research was also based on a positive finding by Bara and his coworkers when they found that the addition of RTIL in poly(RTIL) has increased the gas permeability. This is due to that more rapid gas diffusion occurred as the free volume of membrane increased when RTIL was added [72].

On the other hand, Hudiono has used the RTIL to increase the membrane permeability and also to act as an aid for better interaction between the poly(RTIL) and zeolite (SAPO-34). The result was promising as the permeability of given gases like CO_2, N_2, and CH_4 increased accordingly. However, the selectivity was slightly decrease as they claimed that the RTIL used which is emim[Tf_2N] was not selective towards CO_2/CH_4 separation [71]. Nonetheless, the result proved that the addition of RTIL could increase the polymer-zeolite adhesion in MMM as RTIL also acts as the wetting agent for the zeolite.

Hudiono again repeated the same experiment fabricating a three-component mixed matrix membrane but by varying the composition of RTIL and zeolite added in order to determine the optimum condition for the membrane. The CO_2 permeability seems to rise with the increasing amount of RTIL. The CO_2/CH_4 selectivity of the MMM also improved with the presence of SAPO-34 compared to neat poly(RTIL)-RTIL membrane as long as there is sufficient amount of RTIL as the wetting agent. Besides, the team also conducted an investigation of the separation performance by using the vinyl-based poly(RTIL). The addition of RTIL is not essential as they are structurally similar [73].

In contrast, a ternary MMM has been fabricated by Oral and his coworkers by using different materials. The project study on the effect of different RTIL loadings which are emim[Tf_2N] and emim[CF_3SO_3] towards MMM composed of polyimide-zeolite (SAPO-34). The addition of emim[Tf_2N] has performed as expected when the permeability of CO_2 increased while the incorporation of emim[CF_3SO_3] has increased the CO_2/CH_4 selectivity since emim[CF_3SO_3] is selective towards CO_2/CH_4 [74].

CONCLUSIONS

The escalating research in the membrane fabrication for gas separation applications signifies that membranes technology is currently growing and becoming the major focus for industrial gas separation processes. Latest research area using mixed matrix membranes combines the flexibility and low capital cost with improving selectivity, permeability, chemical, thermal, and mechanical strength. Material selection and method of preparation are the most important part in fabricating a membrane. So the next research must be very careful in determining the materials for gas separation and methods applied in the fabrication stage. Even the synthesized MMMs were only tested in a small scale, the research of MMMs is worth to be further explored since MMMs have shown better separation performance compared to polymeric and inorganic membranes.

REFERENCES

1. Soregraph, Key World Energy Statistic, The International Energy Agency, 2010.

2. Longterm Outlook to 2030, Natural Gas Demand and Supply, The European Union of The Natural Gas Industry, 2010.

3. "Natural Gas and Environment—Emission from the Combustion of Natural Gas," copyright 2004–2010, http://www.naturalgas. org/environment/naturalgas.asp#emission.

4. A. Wan and A. Rusmidah, Natural Gas, Universiti Teknologi Malaysia, 2010.

5. D. David and D. Kishore, Recent Development in CO_2 Removal Membrane Technology, UOP, 1999.

6. M. I. Fauzi and A. Akkil, Meeting Technical Challenge in Developing High CO_2 Gas Field Offshore, Petronas Carigali Sdn. Bhd., 2008.

7. Fuels Providers, Natural Gas Specs Sheet, The National Petroleum Agency, 2002.

8. Separation Process, Membrane Separation Process, Membrane Properties, 1998.

9. Separation Process, Introduction to Membrane, Chapter 1, 1998.

10. K. Scott, Membrane Separation Technology, Scientific & Technical Information, Oxford, UK, 1990.

11. H. Strathmann, "Membrane separation processes: current relevance and future opportunities," AIChE Journal, vol. 47, no. 5, pp. 1077–1087, 2001.

12. S. Morooka and K. Kusakabe, "Microporous inorganic membranes for gas separation," MRS Bulletin, vol. 24, no. 3, pp. 25–29, 1999.

13. A. F. Ismail and L. I. B. David, "A review on the latest development of carbon membranes for gas separation," Journal of Membrane Science, vol. 193, no. 1, pp. 1–18, 2001

14. W. A. W. Abdul Rahman, "Formation and characterization of mixed matrix composite materials for efficient energy gas

separation," Project Report, Faculty of Chemical and Natural Resources Engineering, Universiti Teknologi Malaysia, 2006.

15. J. A. Ritter and A. D. Ebner, "Carbon dioxide separation technology—R&D needs for the chemical and petrochemical industries," Chemical Industry Vision 2020, 2007.

16. T. Visser and M. Wessling, "When do sorption-induced relaxations in glassy polymers set in?"Macromolecules, vol. 40, no. 14, pp. 4992–5000, 2007.

17. A. Bos, I. G. M. Pünt, M. Wessling, and H. Strathmann, "CO_2-induced plasticization phenomena in glassy polymers," Journal of Membrane Science, vol. 155, no. 1, pp. 67–78, 1999.

18. J. D. Wind, D. R. Paul, and W. J. Koros, "Natural gas permeation in polyimide membranes," Journal of Membrane Science, vol. 228, no. 2, pp. 227–236, 2004.

19. J. D. Wind, S. M. Sirard, D. R. Paul, P. F. Green, K. P. Johnston, and W. J. Koros, "Relaxation dynamics of CO_2 diffusion, sorption, and polymer swelling for plasticized polyimide membranes,"Macromolecules, vol. 36, no. 17, pp. 6442–6448, 2003.

20. A. J. Bird and D. L. Trimm, "Carbon molecular sieves used in gas separation membranes," Carbon, vol. 21, no. 3, pp. 177–180, 1983.

21. T. H. Kim, W. J. Koros, G. R. Husk, and K. C. O'Brien, "Relationship between gas separation properties and chemical structure in a series of aromatic polyimides," Journal of Membrane Science, vol. 37, no. 1, pp. 45–62, 1988.

22. J. S. McHattie, W. J. Koros, and D. R. Paul, "Effect of isopropylidene replacement on gas transport properties of polycarbonates," Journal of Polymer Science B, vol. 29, no. 6, pp. 731–746, 1991

23. C. L. Aitken, W. J. Koros, and D. R. Paul, "Gas transport properties of biphenol polysulfones,"Macromolecules, vol. 25, no. 14, pp. 3651–3658, 1992.

24. L. A. Pessan and W. J. Koros, "Isomer effects on transport properties of polyesters based on bisphenol-A," Journal of Polymer Science B, vol. 31, no. 9, pp. 1245–1252, 1993.

25. D. R. B. Walker and W. J. Koros, "Transport characterization of a polypyrrolone for gas separations,"Journal of Membrane Science,

vol. 55, no. 1-2, pp. 99–117, 1991.

26. X. Gao, Z. Tan, and F. Lu, "Gas permeation properties of some polypyrrolones," Journal of Membrane Science, vol. 88, no. 1, pp. 37–45, 1994.

27. J. S. McHattie, W. J. Koros, and D. R. Paul, "Gas transport properties of polysulphones: 2. Effect of bisphenol connector groups," Polymer, vol. 32, no. 14, pp. 2618–2625, 1991.

28. Y. Liu, T. S. Chung, R. Wang, D. F. Li, and M. L. Chng, "Chemical cross-linking modification of polyimide/poly(ether sulfone) dual-layer hollow-fiber membranes for gas separation," Industrial and Engineering Chemistry Research, vol. 42, no. 6, pp. 1190–1195, 2003.

29. G. Xomeritakis, C. Y. Tsai, and C. J. Brinker, "Microporous sol-gel derived aminosilicate membrane for enhanced carbon dioxide separation," Separation and Purification Technology, vol. 42, no. 3, pp. 249–257, 2005.

30. L. Zhang, K. E. Gilbert, R. M. Baldwin, and J. Douglas Way, "Preparation and testing of carbon/silicalite-1 composite membranes," Chemical Engineering Communications, vol. 191, no. 5, pp. 665–681, 2005.

31. M. P. Bernal, J. Coronas, M. Menéndez, and J. Santamaría, "On the effect of morphological features on the properties of MFI zeolite membranes," Microporous and Mesoporous Materials, vol. 60, no. 1-3, pp. 99–110, 2003.

32. C. Y. Tsai, S. Y. Tam, Y. Lu, and C. J. Brinker, "Dual-layer asymmetric microporous silica membranes," Journal of Membrane Science, vol. 169, no. 2, pp. 255–268, 2000.

33. R. W. Baker, E. L. Cussler, W. Eykamp, W. J. Koros, R. L. Riley, and H. Strathmann, Membrane Separation Systems—Recent Developments and Future Directions, Noyes Data Corporation, 1991.

34. D. E. W. Vaughan, "The synthesis and manufacture of zeolites," Chemical Engineering Progress, vol. 84, no. 2, pp. 25–31, 1988.

35. M. A. Carreon, Novel Membranes for Efficient CO_2 Separation, University of Lousville, 2011.

36. S. Kulprathipanja, R. W. Neuzil, and N. N. Li, "Separation of fluids by means of mixed matrix membranes in gas permeation,"

US Patent 4,740,219, 1988.

37. T. M. Gür, "Permselectivity of zeolite filled polysulfone gas separation membranes," Journal of Membrane Science, vol. 93, no. 3, pp. 283–289, 1994.

38. L. Yi, Development of Mixed Matrix Membrane for Gas Separation Application, Tsinghua University, 2006.

39. C. M. Zimmerman, A. Singh, and W. J. Koros, "Tailoring mixed matrix composite membranes for gas separations," Journal of Membrane Science, vol. 137, no. 1-2, pp. 145–154, 1997.

40. R. Mahajan, C. Zimmerman, and W. Koros, Fundamental, Practical Aspects of Mixed Matrix Gas Separation Membranes, ACS Symposium Series, 1999.

41. V. Bhardwaj, A. MacIntosh, I. D. Sharpe, S. A. Gordeyev, and S. J. Shilton, "Polysulfone hollow fiber gas separation membranes filled with submicron particles," Annals of the New York Academy of Sciences, vol. 984, pp. 318–328, 2003.

42. R. Mahajan, R. Burns, M. Schaeffer, and W. J. Koros, "Challenges in forming successful mixed matrix membranes with rigid polymeric materials," Journal of Applied Polymer Science, vol. 86, no. 4, pp. 881–890, 2002.

43. M. G. Süer, N. Baç, and L. Yilmaz, "Gas permeation characteristics of polymer-zeolite mixed matrix membranes," Journal of Membrane Science, vol. 91, no. 1-2, pp. 77–86, 1994.

44. D. R. Paul and D. R. Kemp, "The diffusion time lag in polymer membrane containing adsorptive fillers," Journal of Polymer Science C, no. 41, pp. 79–93, 1973.

45. J. M. Duval, B. Folkers, M. H. V. Mulder, G. Desgrandchampsb, and C. A. Smolders, "Adsorbent filled membranes for gas separation. Part 1. Improvement of the gas separation properties of polymeric membranes by incorporation of microporous adsorbents," Journal of Membrane Science, vol. 80, no. 1, pp. 189–198, 1992.

46. S. Kulprathipanja, "Review of recent progress in mixed matrix membranes," Membrane Technology, vol. 105, pp. 6–8, 2000.

47. R. Mahajan and W. J. Koros, "Factors controlling successful formation of mixed-matrix gas separation materials," Industrial

and Engineering Chemistry Research, vol. 39, no. 8, pp. 2692–2696, 2000.

48. J. M. Duval, Adsorbent filled polymeric membranes [Ph.D. thesis], The University of Twente, 1995.

49. Z. Huang, J. F. Su, X. Q. Su, Y. H. Guo, L. J. Teng, and C. M. Yang, "Preparation and permeation characterization of β-zeolite-incorporated composite membranes," Journal of Applied Polymer Science, vol. 112, no. 1, pp. 9–18, 2009.

50. D. Q. Vu, W. J. Koros, and S. J. Miller, "Mixed matrix membranes using carbon molecular sieves: I. Preparation and experimental results," Journal of Membrane Science, vol. 211, no. 2, pp. 311–334, 2003.

51. M. D. Jia, K. V. Peinemann, and R. D. Behling, "Preparation and characterization of thin-film zeolite-PDMS composite membranes," Journal of Membrane Science, vol. 73, no. 2-3, pp. 119–128, 1992.

52. T. W. Pechar, S. Kim, B. Vaughan et al., "Preparation and characterization of a poly(imide siloxane) and zeolite L mixed matrix membrane," Journal of Membrane Science, vol. 277, no. 1-2, pp. 210–218, 2006.

53. Z. Huang, Y. Li, R. Wen, M. M. Teoh, and S. Kulprathipanja, "Enhanced gas separation properties by using nanostructured PES-zeolite 4A mixed matrix membranes," Journal of Applied Polymer Science, vol. 101, no. 6, pp. 3800–3805, 2006.

54. H. H. Yong, H. C. Park, Y. S. Kang, J. Won, and W. N. Kim, "Zeolite-filled polyimide membrane containing 2,4,6-triaminopyrimidine," Journal of Membrane Science, vol. 188, no. 2, pp. 151–163, 2001.

55. D. Sen, Polycarbonate based zeolite 4A filled mixed matrix membranes: preparation, characterization and gas separation performances [Ph.D. thesis], Middle East Technical University, 2008.

56. D. R. Paul and D. R. Kemp, "Diffusion time lag in polymer membranes containing adsorptive fillers," Journal of Polymer Science C, no. 41, pp. 79–93, 1973.

57. J. D. Figueroa, T. Fout, S. Plasynski, H. McIlvried, and R. D. Srivastava, "Advances in CO_2 capture technology-The U.S.

Department of Energy's Carbon Sequestration Program," International Journal of Greenhouse Gas Control, vol. 2, no. 1, pp. 9–20, 2008.

58. M. Smiglak, W. M. Reichert, J. D. Holbrey et al., "Combustible ionic liquids by design: is laboratory safety another ionic liquid myth?" Chemical Communications, no. 24, pp. 2554–2556, 2006.

59. M. J. Earle, J. M. S. S. Esperança, M. A. Gilea et al., "The distillation and volatility of ionic liquids,"Nature, vol. 439, no. 7078, pp. 831–834, 2006.

60. D. Camper, J. Bara, C. Koval, and R. Noble, "Bulk-fluid solubility and membrane feasibility of Rmim-based room-temperature ionic liquids," Industrial and Engineering Chemistry Research, vol. 45, no. 18, pp. 6279–6283, 2006.

61. T. K. Carlisle, J. E. Bara, C. J. Gabriel, R. D. Noble, and D. L. Gin, "Interpretation of CO_2 solubility and selectivity in nitrile-functionalized room-temperature ionic liquids using a group contribution approach," Industrial and Engineering Chemistry Research, vol. 47, no. 18, pp. 7005–7012, 2008.

62. A. Finotello, J. E. Bara, D. Camper, and R. D. Noble, "Room-temperature ionic liquids: temperature dependence of gas solubility selectivity," Industrial and Engineering Chemistry Research, vol. 47, no. 10, pp. 3453–3459, 2008.

63. Y. I. Park, B. S. Kim, Y. H. Byun, S. H. Lee, E. W. Lee, and J. M. Lee, "Preparation of supported ionic liquid membranes (SILMs) for the removal of acidic gases from crude natural gas," Desalination, vol. 236, no. 1-3, pp. 342–348, 2009.

64. D. Camper, C. Becker, C. Koval, and R. Noble, "Low pressure hydrocarbon solubility in room temperature ionic liquids containing imidazolium rings interpreted using regular solution theory,"Industrial and Engineering Chemistry Research, vol. 44, no. 6, pp. 1928–1933, 2005.

65. P. Scovazzo, J. Kieft, D. A. Finan, C. Koval, D. DuBois, and R. Noble, "Gas separations using non-hexafluorophosphate [PF6]-anion supported ionic liquid membranes," Journal of Membrane Science, vol. 238, no. 1-2, pp. 57–63, 2004.

66. H. Ohno, M. Yoshizawa, and W. Ogihara, "Development of new class of ion conductive polymers based on ionic liquids," Electrochimica Acta, vol. 50, no. 2-3, pp. 255–261, 2004.

67. X. Hu, J. Tang, A. Blasig, Y. Shen, and M. Radosz, "CO_2 permeability, diffusivity and solubility in polyethylene glycol-grafted polyionic membranes and their CO_2 selectivity relative to methane and nitrogen," Journal of Membrane Science, vol. 281, no. 1-2, pp. 130–138, 2006.

68. J. E. Bara, S. Lessmann, C. J. Gabriel, E. S. Hatakeyama, R. D. Noble, and D. L. Gin, "Synthesis and performance of polymerizable room-temperature ionic liquids as gas separation membranes,"Industrial and Engineering Chemistry Research, vol. 46, no. 16, pp. 5397–5404, 2007.

69. J. Tang, W. Sun, H. Tang, M. Radosz, and Y. Shen, "Enhanced CO_2 absorption of poly(ionic liquid)s,"Macromolecules, vol. 38, no. 6, pp. 2037–2039, 2005.

70. J. E. Bara, C. J. Gabriel, S. Lessmann et al., "Enhanced CO_2 separation selectivity in oligo(ethylene glycol) functionalized room-temperature ionic liquids," Industrial and Engineering Chemistry Research, vol. 46, no. 16, pp. 5380–5386, 2007.

71. Y. C. Hudiono, T. K. Carlisle, J. E. Bara, Y. Zhang, D. L. Gin, and R. D. Noble, "A three-component mixed-matrix membrane with enhanced CO_2 separation properties based on zeolites and ionic liquid materials," Journal of Membrane Science, vol. 350, no. 1-2, pp. 117–123, 2010.

72. J. E. Bara, D. L. Gin, and R. D. Noble, "Effect of anion on gas separation performance of polymer-room-temperature ionic liquid composite membranes," Industrial and Engineering Chemistry Research, vol. 47, no. 24, pp. 9919–9924, 2008.

73. Y. C. Hudiono, T. K. Carlisle, A. L. LaFrate, D. L. Gin, and R. D. Noble, "Novel mixed matrix membranes based on polymerizable room-temperature ionic liquids and SAPO-34 particles to improve CO_2 separation," Journal of Membrane Science, vol. 370, no. 1-2, pp. 141–148, 2011.

74. C. A. Oral, R. D. Noble, and S. B. Tantekin-Ersolmaz, "Ternary mixed-matrix membranes containing room temperature ionic liquids," in Proceedings of the North American Membrane Society Conference (NAMS '11), 2011.

Chapter 2

Combustion of Syngas Fuel in Gas Turbine Can Combustor

Chaouki Ghenai

Department of Ocean and Mechanical Engineering, College of Engineering and Computer Science, Florida Atlantic University, 777 Glades Road, 36-177, Boca Raton, FL 33134, USA

ABSTRACT

Numerical investigation of the combustion of syngas fuel mixture in gas turbine can combustor is presented in this paper. The objective is to understand the impact of the variability in the alternative fuel composition and heating value on combustion performance and emissions. The gas turbine can combustor is designed to burn the fuel efficiently, reduce the emissions, and lower the wall temperature. Syngas mixtures with different fuel compositions are produced through different coal and biomass gasification process technologies. The composition of the fuel burned in can combustor was changed from natural gas (methane) to syngas fuel with hydrogen to carbon monoxide (H_2/CO) volume ratio ranging from 0.63 to 2.36. The mathematical models used for syngas fuel combustion consist of the k- model for turbulent flow, mixture fractions/PDF model for no premixed gas combustion, and P-1 radiation model. The effect of syngas fuel composition and lower heating value on the flame shape, gas temperature, mass of carbon dioxide (CO_2) and nitrogen oxides (NO_x) per unit of energy generation is presented in this paper. The results obtained in this study show the change in gas turbine can combustor performance with the same power generation when natural gas or methane fuel is replaced by syngas fuels.

INTRODUCTION

Over the past decades domestic and imported oil was used for transportation, and domestic coal and natural gas have been used as the primary fuels for power generation systems. Today the emission regulations for power plant have become more stringent. The concern today with the combustion of fossil fuels is the new emission regulations for power plant with regards to carbon dioxides (CO_2) and nitrogen oxides (NO_x). Nitrogen oxides (NO_x) are responsible for smog and acid rain, and the carbon dioxides (CO_2) are one of the main greenhouse gases responsible for global warming. Another concern with fossil fuels is the high cost of imported oil. Alternative fuels that can be produced using local feed stocks, burn efficiently, and produce low emissions are needed. With the development of advanced technologies, coal, biomass, or waste products can be used in power generation systems to produce low emissions comparable to the ones obtained with natural gas fuel. This can be achieved through the Integrated Gasification Combined Cycle (IGCC). Different types of gasifiers are used to gasify the solid fuels (coal, biomass or waste products) and produce synthetic gas. The syngas is then cleaned and burned in gas turbine and the hot exhaust gas is used to produce steam for steam turbine. Future power generation systems using Integrated Gasification Combined Cycle (IGCC) will have a higher efficiency and generate lower carbon dioxide and nitrogen oxides emissions. The IGCC systems are used to produce syngas fuel with different compositions from solid fuel feed stocks using different gasification process technologies. The syngas produced through the gasification process consists mainly of hydrogen (H_2) and carbon monoxide (CO) and inert gas such as nitrogen (N_2), water vapor (H_2O), and carbon dioxide (CO_2). Syngas has also less energy density (KJ/Kg) than natural gas. The main characteristics of the syngas fuel are the lower heating value, the H_2/CO ratio, and the fraction (up to 50%) of noncombustible such as steam, carbon dioxide, and nitrogen. For syngas fuels combustion, the effect of hydrogen content is very important. The burning velocity increases with the hydrogen content because the density of the mixture is very low compared to the density of natural gas. The replacement of methane with syngas with high hydrogen content will help to reduce the CO_2 emissions. On the other hand, the dilution of fuel with nitrogen, water, and carbon dioxide reduces the peak flame temperature and consequently the emissions.

Gas turbines are designed primarily to be fueled with natural gas (consisting primarily of methane), and supplying them with syngas (fuel gas from biomass, coal, and waste gasification) presents certain challenges that must be addressed. How can we burn efficiently syngas fuel with different chemical compositions and heating values in gas turbine combustors designed for natural gas? In order to meet these challenges, we need to understand the physical and chemical processes of syngas combustion. Information regarding syngas flame shape, flame speed, gas temperatures and pollutant emissions such as (NO_x)and CO_2 for a range of syngas compositions and heating values is needed for the design of gas turbine combustors. Giles et al. [1] performed a numerical investigation on the effects of syngas composition and diluents on the structure and emission characteristics of syngas counter flow diffusion flame. The counter flow syngas flames were simulated using two representative syngas mixtures, 50%H_2/50%CO and 45%H_2/45%CO/10%CH_4 by volume, and three diluents, N_2, H_2O, and CO_2. The effectiveness of these diluents was characterized in terms of their ability to reduce (NO_x) in syngas flames. The results indicated that syngas no premixed flames are characterized by relatively high temperatures and high (NO_x) concentrations and emission indices. The presence of methane in syngas decreases the peak flame temperature, but increases the formation of prompt NO significantly. They also concluded that the presence of methane in syngas reduces the effectiveness of all three diluents. Lean premixed combustion of hydrogen–syngas/methane fuel mixtures was investigated experimentally by Alavandi and Agarwal [2]. Methane (CH_4) content in the fuel was decreased from 100% to 0% (by volume), with the remaining amount split equally between carbon monoxide (CO) and hydrogen (H_2), the two reactive components of the syngas. Experiments for different fuel mixtures were conducted at a fixed air flow rate, while the fuel flow rate was varied to obtain a range of adiabatic flame temperatures. The CO and nitric oxide (NO_x) emissions were measured downstream of the burner, in the axial direction to identify the post combustion zone and in the transverse direction to quantify combustion uniformity. The results show that increasing H_2/CO content in the fuel mixture decreased both the CO and (NO_x) emissions. Experimental study on the fundamental impact of firing syngas in gas turbines was performed by Oluyede [3]. The goal of this study was to determine the appropriate amount of reduction in firing temperature needed to maintain the same hot section temperatures as experienced with natural gas firing. The results show that volume fraction of hydrogen content in syngas fuel significantly impacts the life of hot sections as a result of higher flame temperature for hydrogen rich fuels and also the moisture content of combustion products. Correlations were obtained indicating the level of firing temperature reduction, necessary for hot section durability in terms of hydrogen contents and lower heating value of the fuel. The combustion of hydrogen-enriched methane in a lean premixed swirl burner was investigated by Schfere [4]. The results, using methane/hydrogen fuel mixtures, showed that the addition of up to 41% hydrogen significantly extended the lean burning limit. For operating conditions near the lean stability limit, the addition of a moderate amount of hydrogen to the methane/air mixture resulted in a significant increase in the OH concentration and a more robust appearing flame. Pater [5] investigated the thermo acoustic instabilities during turbulent syngas combustion. He used numerical method to predict acoustic fields and instabilities during syngas combustion. The model was used to identify frequencies at which instabilities occur.

The challenges of fuel diversity while maintaining superior environmental performance of gas turbine engines were addressed by Rahm et al. [6]. They reviewed the combustion design flexibility that allows the use of a broad spectrum of gas and liquid fuel including emerging synthetic choices. Gases include ultra-low heating value process gas, syngas, ultrahigh hydrogen, or higher heating capability fuels. The integration of heavy-duty gas turbine technology with synthetic fuel gas processes using low value feed stocks in global power generation marketplace was covered by Brdar and Jones [7]. In their paper they summarized the experience gained from several syngas projects and lessons learned that continue to foster cost reductions and improve the operational reliability of gas turbine. They concluded that further improvements in system performance and plant design are needed in the future. The design of combustion systems using syngas as fuel can take advantage of CFD analysis to optimize the efficiency of the combustion system with respect to the limitations of pollutants emission. The aim of this work is to analyze the fundamental impacts of firing syngas in gas turbine combustor and predict the changes in the firing temperature and emissions with respect to natural gas or methane combustion.

GOVERNING EQUATIONS

The mathematical equations describing the syngas fuel combustion are based on the equations of conservation of mass, momentum, and energy together with other supplementary equations for the turbulence and combustion. The standard k-ε turbulence model is used in this study. The equations for the turbulent kinetic energy k and the dissipation rate of the turbulent kinetic energy ε are solved. For non-premixed combustion modeling, the mixture fraction/PDF model is used. The time averaged gas phase equations for steady turbulent flow are:

$$\frac{\partial}{\partial x_j}(\rho u_i \Phi) = -\frac{\partial}{\partial x_i}\left(\Gamma_\Phi \frac{\partial \phi}{\partial x_i}\right) + S_\Phi.$$

(1)

Φ is the dependent variable that can represent the velocity u_i, T the temperature, k the turbulent kinetic energy, ε the dissipation rate of the turbulent kinetic energy, and f the mixture fraction. The governing equations are:

Continuity

$$\frac{\partial \rho u_i}{\partial x_i} = 0.$$

(2)

Momentum Equation

$$\frac{\partial \left(\rho u_i u_j \right)}{\partial x_j} = -\frac{\partial \bar{P}}{\partial x_i} + \frac{\partial \left(\bar{t}_{ij} + \overline{\tau}_{ij} \right)}{\partial x_j},$$

(3)

Where $\bar{t}_{ij}$ is the viscous stress tensor defined as:

$$\bar{t}_{ij} = \mu \left[\left(\frac{\partial \overline{u}_i}{\partial x_j} + \frac{\partial \overline{u}_j}{\partial x_i} \right) - \frac{2}{3} \frac{\partial \overline{u}_k}{\partial x_k} \delta_{ij} \right],$$

$$\delta_{ij} = 1 \quad \text{if} \quad i = j, \quad \delta_{ij} = 0 \quad \text{if} \quad i \neq j.$$

(4)

$\overline{\tau}_{ij}$ is the average Reynolds stress tensor defined as: $\overline{\tau}_{ij} = -\rho \overline{u'_i u'_j}$

$$\overline{\tau}_{ij} = \mu_t \left[\left(\frac{\partial \overline{u}_i}{\partial x_j} + \frac{\partial \overline{u}_j}{\partial x_i} \right) - \frac{2}{3} \frac{\partial \overline{u}_k}{\partial x_k} \delta_{ij} \right] - \frac{2}{3} \left(\overline{\rho k \delta_{ij}} \right),$$

(5)

Where $\bar{k}$ is the average turbulent kinetic energy defined as: $\bar{k} = (1/2)\overline{u'_i u'_j}$, μ_t is the turbulent eddy viscosity expressed as: $\mu_t = c\mu \rho k^2 / \varepsilon$, where C_μ is constant ($C_\mu = 0.09$) and ε is the average dissipation rate of the turbulent kinetic energy and defined as follows: $\bar{\varepsilon} = v\overline{\partial u'_i / \partial x_j \partial u'_i / \partial x_j}$

.

Turbulent Kinetic Energy Equation

$$\frac{\partial \left(\overline{\rho k u_j} \right)}{\partial x_j} = \frac{\partial \left[\left(\mu + \mu_t/\sigma_k \right) \left(\partial \overline{k}/\partial x_j \right) \right]}{\partial x_j} + G_k - \overline{\rho \varepsilon},$$

(6)

Where $\sigma_k = 1$ and G_k is the production of the turbulent kinetic energy defined as

$$G_K = \mu_t \left| \left(\frac{\partial \overline{u}_i}{\partial x_j} + \frac{\partial \overline{u}_j}{\partial x_i} \right) \right| \frac{\partial \overline{u}_i}{\partial x_j} - \frac{2}{3} \frac{\partial \overline{u}_i}{\partial x_j} \delta_{ij} \left[\mu_t \frac{\partial \overline{u}_k}{\partial x_k} + \overline{\rho k} \right].$$

(7)

Dissipation of the Kinetic Energy

$$\frac{\partial \left(\overline{\rho \varepsilon u_j} \right)}{\partial x_j} = C_{\varepsilon 1} \frac{\overline{\varepsilon}}{k} G_k + \frac{\partial \left[\left(\mu + \mu_t/\sigma_\varepsilon \right) \left(\partial \overline{\varepsilon}/\partial x_j \right) \right]}{\partial x_j} - C_{\varepsilon 2} \overline{\rho} \frac{\overline{\varepsilon^2}}{k},$$

(8)

Where $C_{\varepsilon 1} = 1.44$, $C_{\varepsilon 2} = 1.92$, and $\sigma_\varepsilon = 1.3$.

Mixture Fraction f

In non-premixed combustion, fuel and oxidizer enter the reaction zone in distinct streams. The PDF/mixture fraction model is used for non-premixed combustion modeling. In this approach individual species transport equations are not solved. Instead, equation for the conserved scalar (f) is solved, and individual component concentrations are derived from the predicted mixture fraction distribution. The mixture fraction equation is given by

$$\frac{\partial \left(\overline{\rho f u_j} \right)}{\partial x_j} = \frac{\partial \left[\left(\mu_t/\sigma_t \right) \left(\partial \overline{f}/\partial x_j \right) \right]}{\partial x_j} + S_m.$$

(9)

The mixture fraction, f can be written in terms of elemental mass fraction as

$$f = \frac{Z_k - Z_{k,O}}{Z_{k,F} - Z_{k,O}},$$

(10)

Where Z_k is the element mass fraction of some element k. Subscripts F and O denote fuel and oxidizer inlet stream values, respectively. For the mixture fraction approach, the equilibrium chemistry PDF model is used. The equilibrium system consists of 13 species (C, CH_4, CO, CO2, H, H_2, H_2O, N_2, NO, O, O_2, OH, HO_2). The chemistry is assumed to be fast enough to achieve equilibrium.

Energy Equation

$$\frac{\partial\left(\left(\rho E + p\right)uj\right)}{\partial x_j}$$

$$= \frac{\partial\left[(k_{eff})\left(\partial \overline{T}/\partial x_j\right) - \sum_j h_j J_j + \left(\overline{\tau}_{eff} u_j\right)\right]}{\partial x_j} + S_h,$$

(11)

Where E is the total energy ($E = h - p/\rho + v2/2$, where h is the sensible enthalpy), k_{eff} is the effective conductivity ($k + k_t$: laminar and turbulent thermal conductivity), J_j is the diffusion flux of species j, and S_h is the term source that includes the heat of chemical reaction, radiation, and any other volumetric heat sources.

Equation for the P-1 Radiation Model—Radiation Flux Equation

The P-1 radiation model is used in this study to simulate the radiation from the flame. The radiation model is based on the expansion of the radiation intensity into an orthogonal series of spherical harmonics (Cheng [8] and Siegel and Howell [9]). The P-1 radiation model is the simplest case of the P-N model. If only four terms in the series are used, the following equation is obtained for the radiation flux:

$$q_r = -\frac{1}{3(a + \sigma_S) - C\sigma_S} \nabla G,$$

(12)

Where a is the absorption coefficient, σ_S is the scattering coefficient, G is the incident radiation, and C is the linear anisotropic phase function coefficient (Cheng [8] and Siegel and Howell [9]).

The transport equation for G is

$$\nabla \cdot (\Gamma \nabla G) - aG + 4a\sigma T^4 = S_G,$$

$$\Gamma = \frac{1}{(3(a + \sigma_S) - C\sigma_S)}.$$

(13)

GEOMETRY, BOUNDARY CONDITIONS, MESH, AND NUMERICALMETHOD

The gas turbine can combustor is designed to burn the fuel efficiently, lower the emissions, and keep the combustor wall temperatures low. The basic geometry of the gas turbine can combustor is shown in Figure 1. The size of the combustor is 590mm in the Z direction, 250mm in the Y direction, and 230mm in the X direction. The primary inlet air is guided by vanes to give the air a swirling velocity component (see Figure 1(a)). The boundary conditions of the primary air are as follows: the injection velocity is 10 m/s, the temperature is 300 K, the turbulence intensity is 10%, mixture fraction $f = 0$ and the injection diameter is 85 mm. The fuel is injected through six fuel inlets in the swirling primary air flow (see Figure 1(a)). The boundary conditions of the fuel are as follows: mass flow rate, 0.001 Kg/s, the temperature is 300 K, the turbulence intensity is 10%, mixture fraction $f = 1$ and the injector diameter is 4.2 mm. The secondary air or dilution air is injected at 0.1 meters from the fuel injector to control the flame temperature and NO_x emissions. The secondary air is injected in the combustion chamber through six side air inlets each with a diameter of 16mm (see Figure 1(b)). The boundary conditions of the secondary air are as follows: the injection velocity is 6m/s, the temperature is 300 K, the turbulence intensity is 10%, mixture fraction $f = 0$ and the injection diameter is 16 mm. The can combustor outlet has a rectangular shape (see Figure 1(b)) with an area of 0.0150m2. A quality mesh was generated for the can combustor (see Figure 2). The mesh consists of 106,651 cells or elements (74189 tetrahedra, 30489 wedges, and 1989 pyramids), 234368 faces, and 31433 nodes. The grid quality was checked, and the results showed a maximum cell squish of 0.94, maximum cell skewness of 0.99, and a maximum aspect ratio of 83.17. The finite volume method and the first-order upwind method were used to solve the governing equations. The solution procedure for a single-mixture-fraction system was to (1) complete the calculation of the PDF look-up tables first, (2) start the reacting flow simulation to determine the flow files and predict the spatial distribution of the mixture fraction, (3) continue the reacting flow simulation until a convergence solution was achieved, and (4) determine the corresponding values of the temperature and individual chemical species mass fractions from the look-up tables. The convergence criteria were set to 10^{-3} for the continuity, momentum, turbulent kinetic energy, dissipation rate of the turbulent kinetic energy, and the mixture fraction. For the energy and the radiation equations, the convergence criteria were set to 10^{-6}.

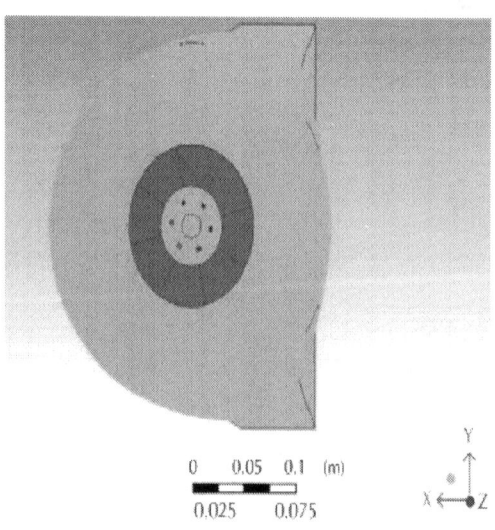

0   0.05   0.1   (m)

0.025   0.075

(a)

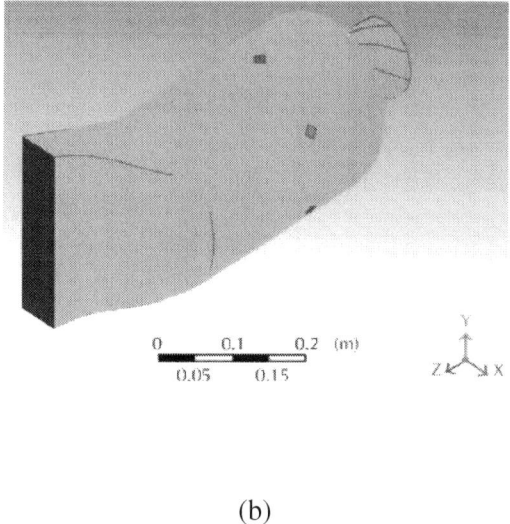

(b)

Figure 1: Geometry of the gas turbine can combustor, (a) primary air (blue) and six fuel inlets, (b) secondary air (blue) and outlet.

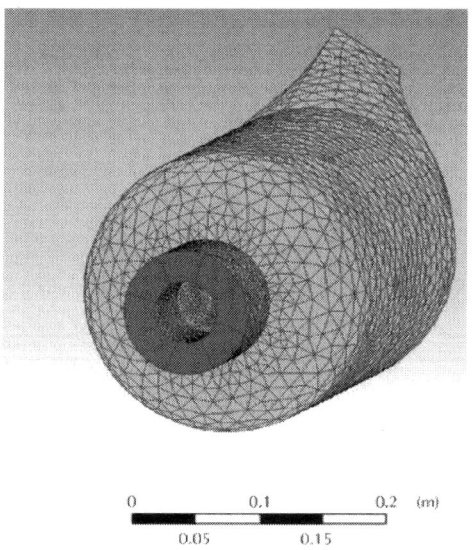

Figure 2: Mesh for the basic geometry of gas turbine can combustor.

RESULTS

The impact of the variability in the syngas fuel composition and low heating value on the combustion performance and emissions in gas turbine can combustor is performed in this study. Table 1 shows the composition for the five syngas fuel and their lower heating values selected for this CFD analysis. The syngas fuels were produced using different gasification processes (Todd [10]) and using different feed stocks (coal, biomass, waste). The range of the constituents volume fractions for the selected syngas fuels are hydrogen (1) $(H_2) = 22.6\%-61.6\%$, (2) carbon monoxide $(CO) = 23.6\%-46.6\%$, (3) methane $(CH_4) = 0.1\%-6.9\%$, (4) carbon dioxide $(CO_2) = 5.6\%-17.9\%$, (5) Nitrogen $(N_2) = 1.1\%-49.3\%$, and (6) water $(H_2O) = 0.3\%-39.8\%$. The hydrogen to carbon monoxide volume ratio for these five syngasis between 0.63 and 2.36. Table 1 shows also that the lower heating values for the syngas fuels are smaller compared to the lower heating value of the methane. It is also noted that syngas 1 (Schwarze Pumpe) has the highest hydrogen volume fraction (61.9%), syngas 2 (Exxon Singapore) has the highest carbon dioxide volume fraction, syngas 3 (Tampa) has the highest carbon monoxide volume fraction (46.6%), and syngas 5 (Sarlux) has the highest water vapor volume fraction (39.8%).

Table 1: Syngas compositions

Constituents	Syngas 1 Schwarze pumpe	Syngas 2 Exxon singapore	Syngas 3 Tampa	Syngas 4 PSI	Syngas 5 Sarlux	Methane (CH_4)
H2	61.9	44.5	37.2	24.8	22.7	0
CO	26.2	35.4	46.6	39.5	30.6	0
CH_4	6.9	0.5	0.1	1.5	0.2	100
CO_2	2.8	17.9	13.3	9.3	5.6	0
N_2	1.8	1.4	2.5	2.3	1.1	0
H_2O	0.4	0.3	0.3	22.6	39.8	0
Volumetric heating value KJ/m³	12492	9477	9962	8224	—	33570
Lower heating value MJ/Kg	27.8	12.8	12.7	10.4	—	50.1
H2/CO	2.36	1.26	0.8	0.63	0.74	—

The contours of the predicted gas temperature for the combustion of methane in gas turbine can combustor are shown in Figures 3 and 4. The maximum gas temperature for methane combustion is 2200 K. For the validation of the combustion model, the predicted flame temperature for methane combustion was compared to the adiabatic flame temperature. For natural gas or methane fuel and with initial atmospheric conditions (1 bar and 20 C), the theoretical flame temperature produced by the flame with a fast combustion reaction is 2233 K. The predicted maximum temperature of the combustion products or the adiabatic flame temperature compares well with the theoretical adiabatic flame temperature. The peak gas temperature is located in the primary reaction zone. The fuel from the six injectors is first mixed in the swirling air before burning in the primary reaction zone. The gas temperature decreases after the primary reaction zone due to the dilution of the flame with the secondary air.

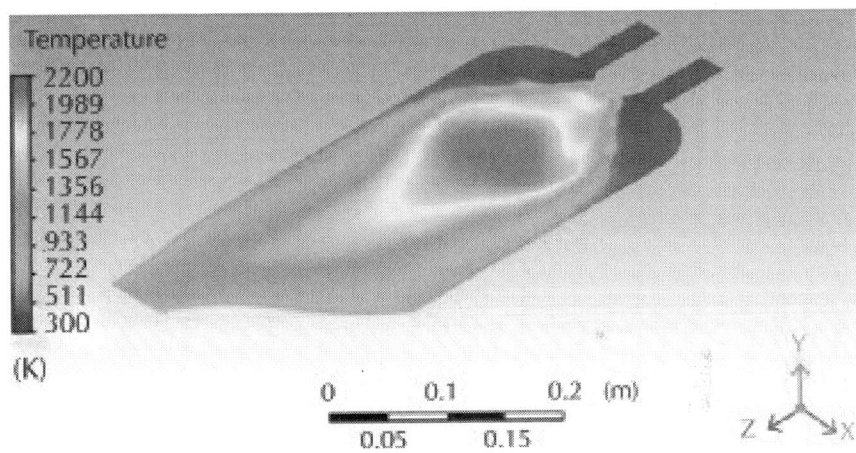

Figure 3: Temperature contours (X-Z Plane, Y = 0): combustion of methane in gas turbine can combustor.

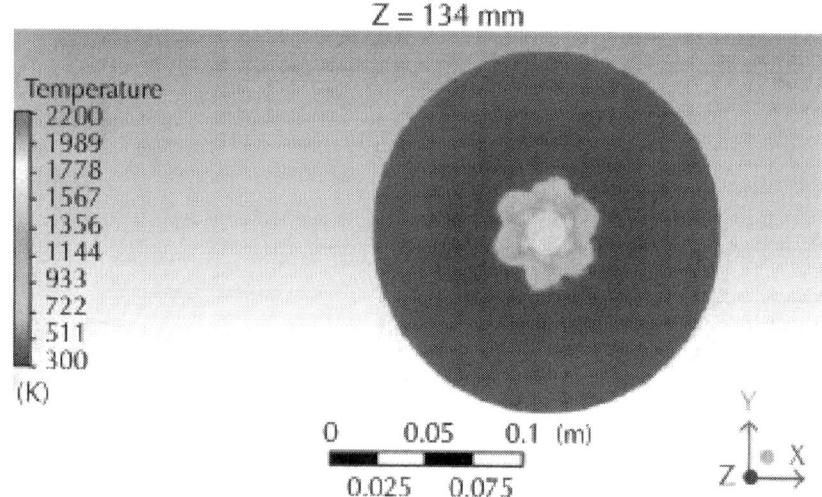

(a)

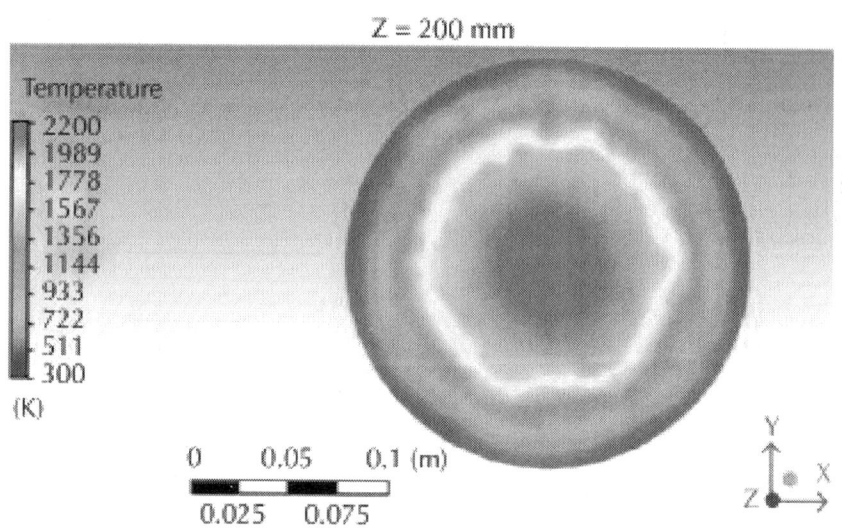

(b)

Z = 300 mm

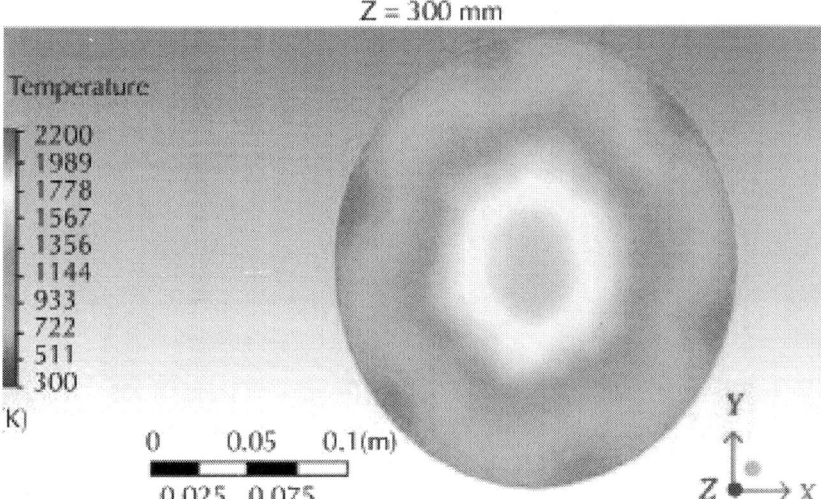

(c)

Z = 400mm

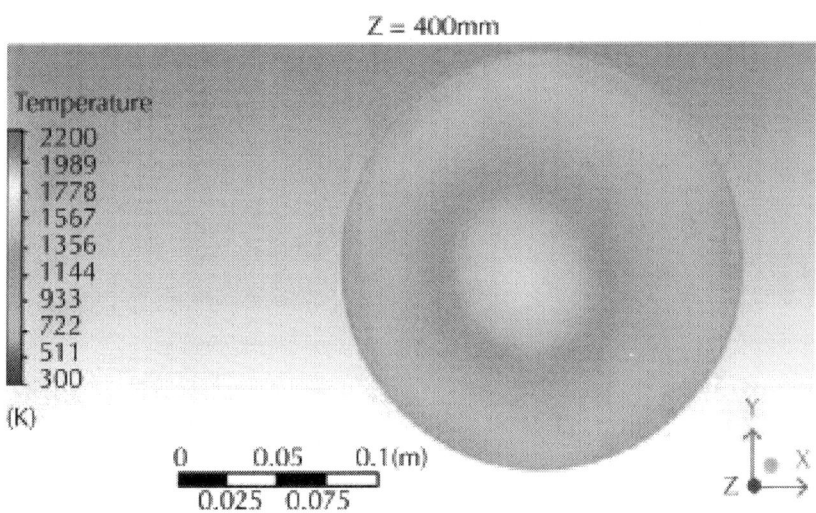

(d)

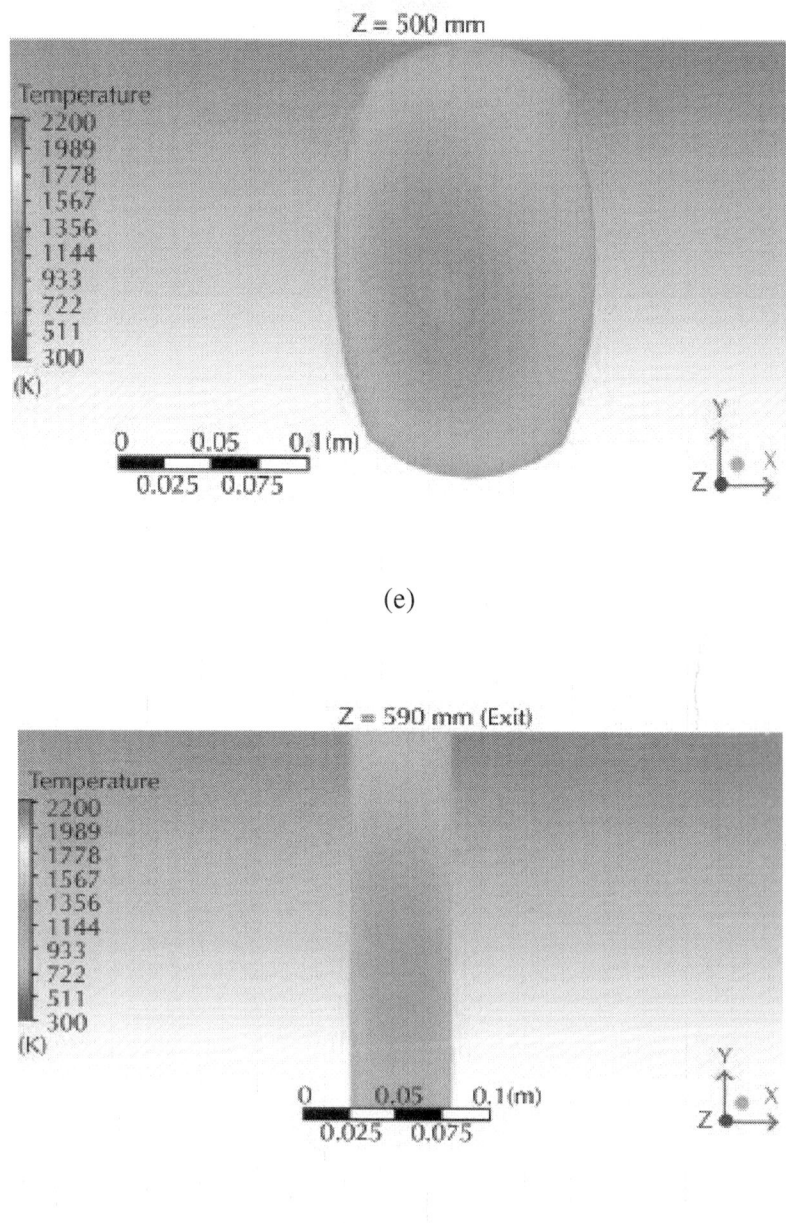

(e)

(f)

Figure 4: Temperature contours (*X-Y* Plane)—Combustion of methane in gas turbine can combustor.

Figure 4 shows the temperature contours (*X-Y* Plane) for methane combustion in can combustor at different axial positions (Z=134 mm to 590 mm). The first contour (Z=134 mm) represents the gas temperature near the six fuel injections. The fuel is injected from the six fuel inlets and mixed with the swirling air before the start of the combustion. The size of the flame increases downstream and reach a maximum radius at Z=200 mm. The radius and temperature of the flame decrease after that with the increase of the axial distance Z (Z=300, 400, and 500 mm). The lowest gas temperature is reached at the exit (Z=590 mm) of the gas turbine can combustor. A uniform gas temperature distribution is obtained at the exit of the can combustor as shown in Figures 3 and 4.

The contours of the velocity w (*Z*-component) for methane combustion in gas turbine can combustor is shown in Figure 5. The primary air is injected in the Z direction with initial velocity of 10 m/s. The primary air is accelerated (up to 16 m/s) at the entrance of the combustors due to the presence of swirlers vanes (see Figure 1). Swirlers curved vanes are used to generate recirculation zone at the entrance of the combustion chamber. Strong recirculation regions are produced in the fuel injection region. This will help to increase the turbulence and mix very well the fuel and air in the primary reaction zone. This will in turn burn the fuel efficiently and reduce the pollutants emissions. Figure 6 shows the contours (-(*X-Y* Plane) of the velocity swirling strength for Z = 134 mm to 400. Strong recirculation regions with strong swirling strength are shown at the entrance regions near the fuel injection. The contours show that the velocity swirling strength decreases downstream with the increase of the axial distance Z. The velocity swirling strength contours at Z = 200 shows an annulus region with high swirling strength.

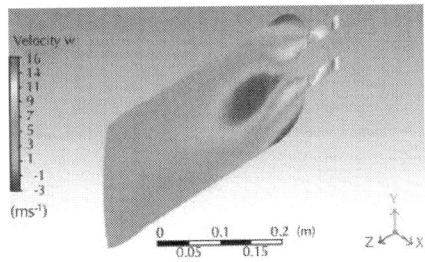

Figure 5: Contours of the velocity w (z-direction) in the y-z plane: combustion of methane in the gas turbine can combustor.

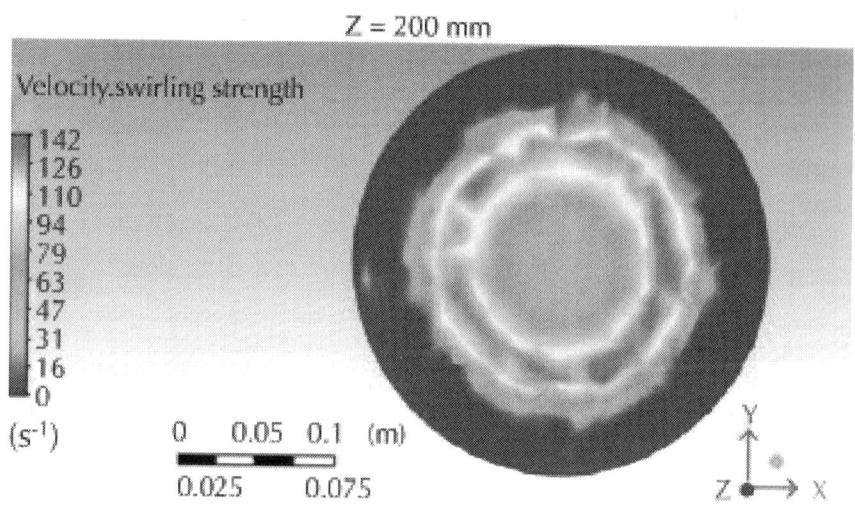

(a)

(b)

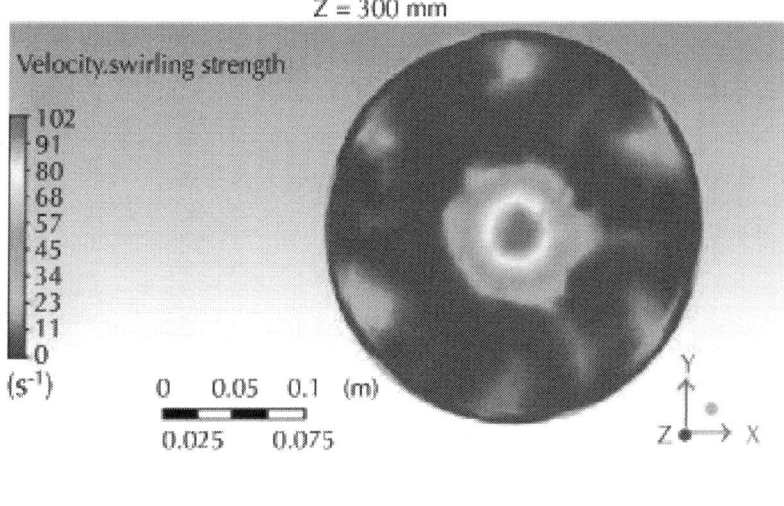

(c)

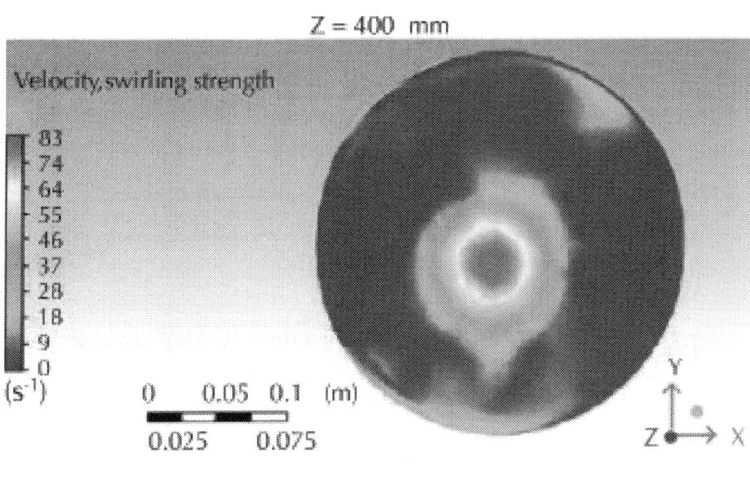

(d)

Figure 6: Contours of velocity swirling strength (X-Y plane): combustion of methane in gas turbine can combustor.

The results of methane combustion in gas turbine can combustor are used as a baseline for comparison with syngas fuel combustion. The effect of syngas fuel compositions on gas temperature, CO_2 and NO_x emissions are investigated first. Table 2 shows the fuel, primary and secondary air injection conditions, and the power generated for methane and syngas fuels. It is noted that the first part of the CFD analysis was performed with the same fuel mass flow rate and different power generation. The gas temperature contours (X-Z plane at Y= 0) for methane and syngas fuels combustion in gas turbine can combustor are shown in Figure 7. The gas temperature for the five syngas tested in this study shows a lower gas temperature compared to the temperature of methane. It is noted that all the syngas fuel have lower heating value than the methane. The volumetric heating value of the methane is 2.68 to 4.08 times higher than the volumetric heating value of the syngas fuels tested in this study. In addition to that the flame temperature depends also on the syngas fuels compositions. For example, the maximum gas temperature predicted for syngas 5 (Sarlux) is about 1734 K. The maximum temperature for syngas 5 is 466 K less than the maximum flame temperature for methane. This is due to the high volume fraction of water (39.8%) in syngas 5. The maximum temperature for syngas 3 is 2008 K compared to 2200 K for methane. Syngas 3 has the highest volume fraction of carbon monoxide CO (46.6%). The syngas fuel composition and heating value are not only affecting the maximum flame temperature but also the shape and size of the flame. The flame shape and size are changing when we switch the fuel from pure methane to syngas fuels. The syngas combustion shows a shorter flame compared to the one obtained with methane. Replacing methane with syngas fuels with different hydrogen concentration will affect the combustion process. The reaction zone or the flame is located in the vicinity of the fuel injection region as shown in Figure 7 for all five syngas with hydrogen to carbon monoxide ratio between 0.63 and 2.36. The flame temperature for syngas gas depends not only on the hydrogen to carbon monoxide ratio (combustible constituents) but also on the volume fraction of the noncombustible (CO_2, N_2, and H_2O) in the syngas. The higher the hydrogen volume fraction in the syngas is, the higher the gas temperature is and the higher the volume fraction of inert gas in syngas is, the lower the gas temperature is. The contours of carbon dioxide (CO_2) mass fraction for methane and syngas fuel (Schwarze Pumpe) combustion are shown in

Figure8. The figure shows higher CO_2 concentrations in the reaction zone. It is also noted that higher CO_2 concentrations are generated with methane when the fuel mass rate was kept constant. With the same fuel flow rate, the power generated by the syngas fuels tested in this study is about 20% to 55% the power generated by the methane. For the same power generation, the fuel mass flow rate of the syngas fuel and the CO_2 emission will be higher. With a constant fuel mass flow rate, the syngas fuel combustion shows less carbon dioxide formation inside the can combustor for all the five syngas tested in this study. The average carbon dioxide mass fractions at the exit of the combustor generated with methane and syngas fuels with constant fuel mass flow rate are shown in Figure 9(a). A net reduction (30% to 49%) of CO_2 emissions at the exit of the combustor when methane is replaced with syngas fuel (see Figure 9(a)) but at the same time the power generated by the syngas gas decreased also by 20% to 55%. The results shown in Figure 9(a) are obtained with methane and syngas fuel with the same mass fuel rate at the injection and different power generated by the can combustor (see Table 2). For the evaluation of the carbon loading to the environment accounted by the can combustor, the results should be presented as the average mass of CO_2 emitted per unit of energy generation. The average mass flow rate of CO_2 emissions (Kg/s) at the exit of the combustor was calculated. The CO_2 mass flow rate (Kg/s) was normalized with power (KW or KJ/s) to determine the mass of CO_2 emitted per unit of energy generation (Kg/KJ). The results of the mass of CO_2 per unit energy generation are shown in Figure 9(b). The results show clearly that mass of CO_2 emitted per unit of energy generation (Kg/KJ) is higher for the syngas fuels (Exxon Singapore, Tampa, and PSI) with lower heating value (10.4–12.8 MJ/Kg) or power generation (10.4 to 12.8 KW). The only syngas fuel that shows a reduction of the average mass of CO_2 emitted per unit of energy generation (Kg/KJ) compared to methane fuel is Schwarze Pumpe syngas. The Schwarze Pumpe syngas fuel ha a lower heating value of 27.8 MJ/Kg, and the power generated during the combustion of this fuel in the can combustor is 27.8 KW. The average mass of CO_2 emitted per unit of energy generation (Kg/KJ) for the Schwarze Pumpe syngas fuel decreased by 12% compared to methane fuel.

Table 2: Fuel and air injection conditions and power generated

Constituents	Syngas 1 Schwarze Pumpe	Syngas 2 Exxon Singapore	Syngas 3 Tampa	Syngas 4 PSI	Syngas 5 Sarlux	Methane (CH$_4$)
Fuel mass flow rate (Kg/s)	0.001	0.001	0.001	0.001	0.001	0.001
Velocity of primary air (m/s)	10	10	10	10	10	10
Velocity of secondary air (m/s)	6	6	6	6	6	6
Total mass flow rate at the exit (Kg/s)	0.0783	0.0783	0.0783	0.0783	0.0783	0.0783
Lower heating value MJ/Kg	27.8	12.8	12.7	10.4	—	50.1
Power KW	27.8	12.8	12.7	10.4	—	50.1

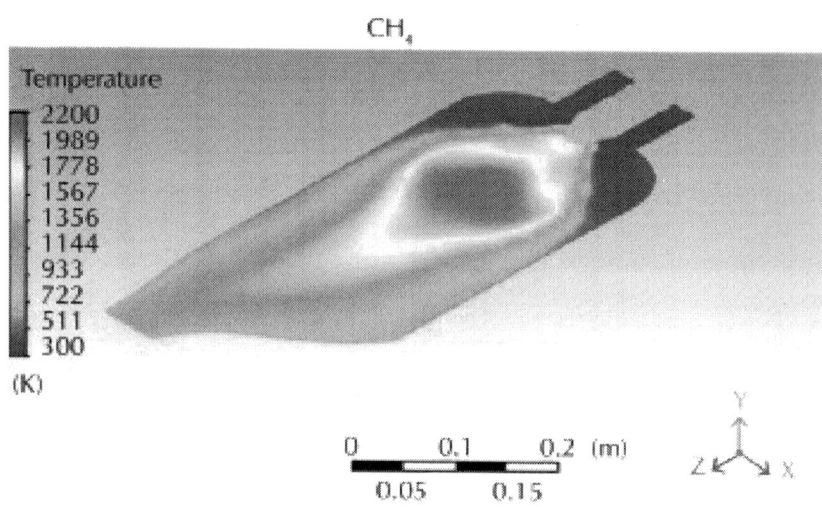

CH$_4$

Temperature

2200
1989
1778
1567
1356
1144
933
722
511
300

(K)

0   0.1   0.2 (m)

0.05   0.15

(a)

Syngas: $H_2/CO = 2.36$

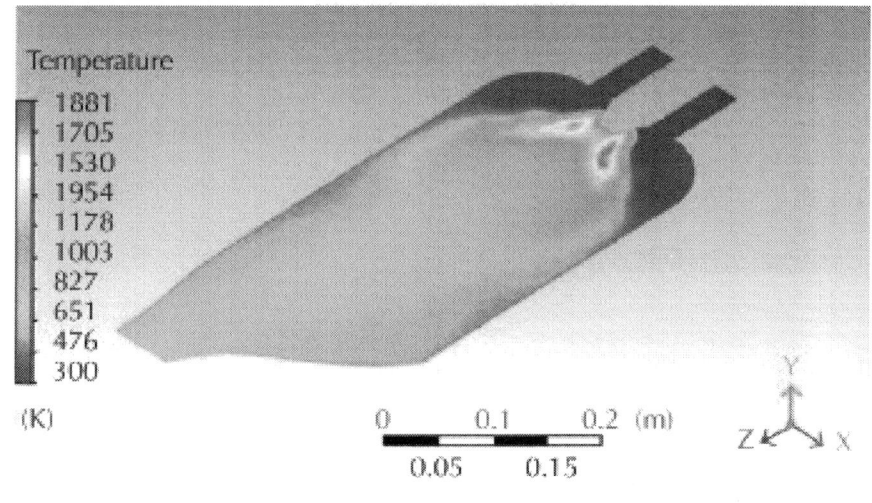

Temperature
1881
1705
1530
1954
1178
1003
827
651
476
300

(K)

0   0.1   0.2 (m)

0.05   0.15

(b)

Syngas: $H_2/CO = 1.26$

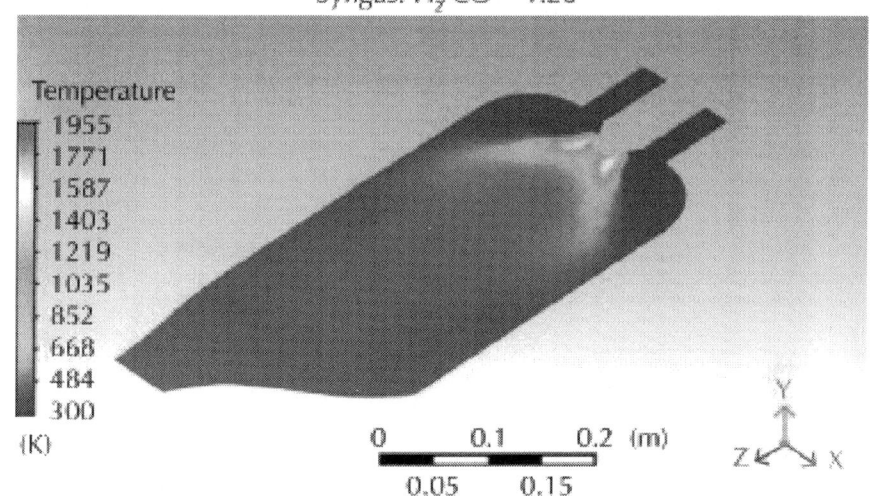

Temperature
1955
1771
1587
1403
1219
1035
852
668
484
300

(K)

0   0.1   0.2 (m)

0.05   0.15

(c)

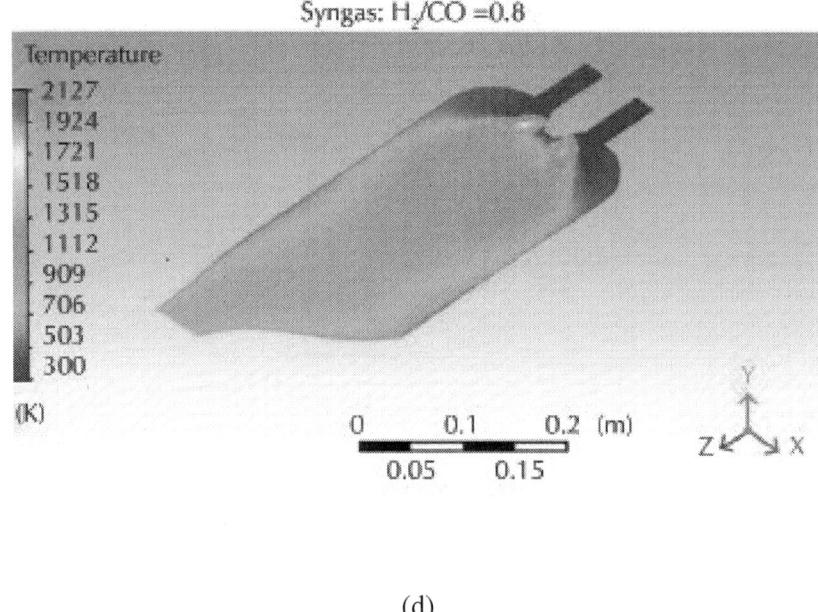

(d)

Figure 7: Temperature contours for natural gas and syngas fuel mixtures with the same fuel mass flow rate: effect of Syngas composition (H_2/CO).

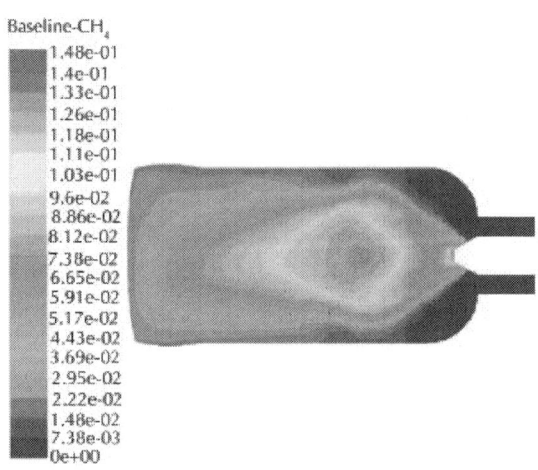

(a)

Syngas 1: Schwarze Pumpe (H_2/CO = 2.36)

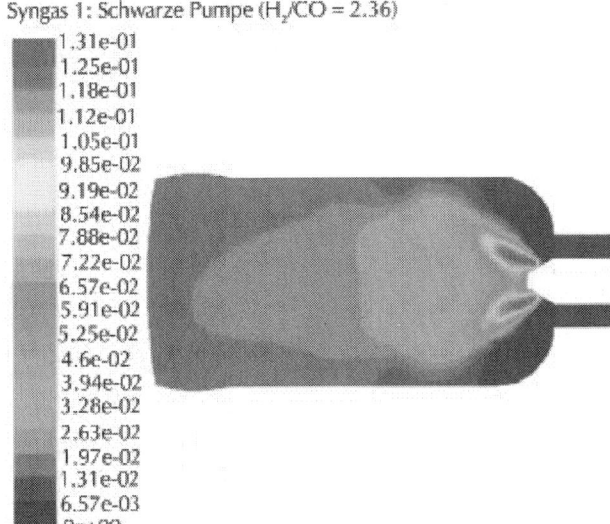

(b)

Figure 8: Contours of CO_2 mass fraction.

Baseline-CH_4

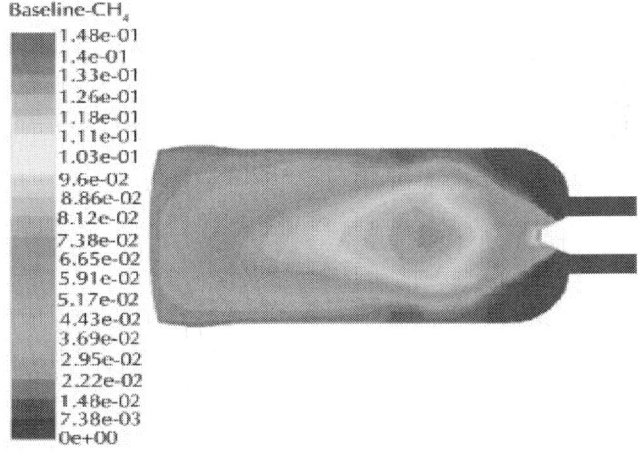

(a)

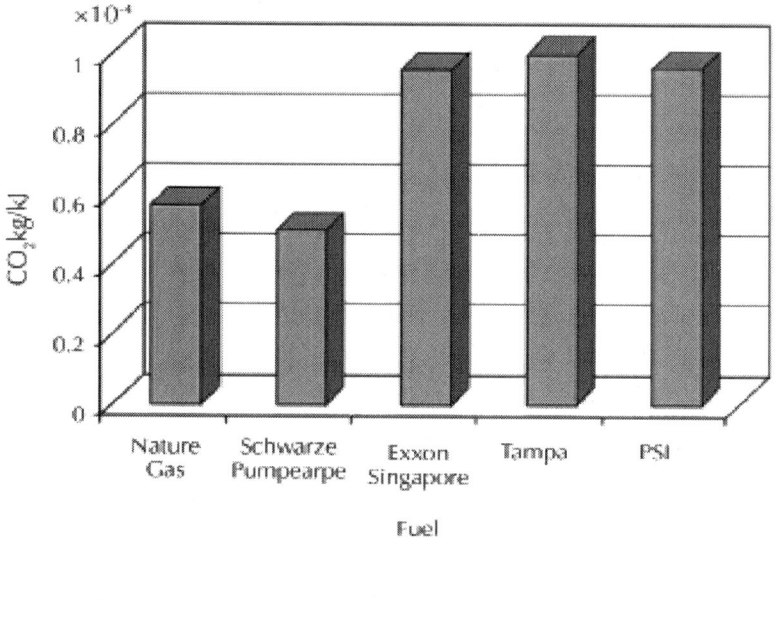

(b)

Figure 9: (a) Average carbon dioxide (CO_2) mass fractions at the exit of can combustor. (b) Average mass of carbon dioxide at the exit per unit of energy generation.

The NO_x emissions from syngas combustion were also calculated in this study. The NO_x concentrations emitted from combustion systems are generally low. The NO_x chemistry has negligible influence on the predicted flow field, temperature, and major combustion product concentrations. The NO_x model used in this study was a postprocessor to the main combustion calculation. First the reacting flows are simulated without NO_x emissions until the convergence of the main combustion calculation was obtained; after that the desired NO_x models (thermal, prompt, fuel) were enabled to predict the NO_x emissions. For NO_x emissions prediction (post processing), only the equation for NO will be solved, and the solution for the other equations (mass, momentum, energy, radiation, turbulent kinetic energy, dissipation of the turbulent kinetic energy, and mixture fraction) will be turned off. The calculation for NO_x emissions will continue until the NO species residuals are below 10^{-6} (convergence of NO solution). The contours of NO mass fractions inside the can combustor obtained with the same fuel mass flow rate or different power (see Table 2) are shown in Figure10. Like the gas temperature contours, the NO mass fraction inside the combustor decreases when the baseline fuel (CH_4) is replaced with syngas fuel with lower heating value. The thermal NO emissions are function of the gas temperature. The higher is the temperature, the higher is the NO mass fractions. The gas temperature during syngas combustion depends on the lower heating value of the fuel and fuel composition. The temperature increases with the increase of the volume fraction of the combustible constituents in the fuel such as hydrogen, carbon monoxides and methane. On the other hand, the presence of non-combustible constituents in the syngas such as water, nitrogen, and carbon dioxides reduces the temperature of the flame and consequently the NO mass fractions. Figure 11(a) shows the average NO mass fraction at the exit of the can combustor. If the fuel mass flow rate is kept constant and the power generated from the syngas fuels is 20% to 55% the power generated by methane fuel, the net NO mass fraction reduction for syngas 1 and syngas 5 is, respectively, 11.5% and 97.6% compared to methane. This reduction is proportional to the amount of inert constituents in syngas fuel. The volume fraction of inert constituents for syngas 1 and syngas 5 is, respectively, 5% and 46.5%. The higher is the amount on noncombustible constituents in the syngas fuel the lower is the NO mass fraction at the exit of the can combustor. For non-premixed combustion, syngas fuel is used to control the NO_x emissions by diluting the syngas with nitrogen, steam, and carbon dioxides. To assess the amount of NO_x loading to the environment accounted by the can combustor, the results should be presented as the average mass of NO emitted per unit of energy generation. The average mass flow rate of NO emissions (Kg/s) at the exit of the combustor was calculated. The NO mass flow rate (Kg/s) was normalized with power (KW or KJ/s) to determine the mass of NO emitted per unit of energy generation (Kg/KJ). The results of the mass of NO per unit energy generation are shown in Figure 11(b). The results show that mass of NO per unit energy generation is higher for most of the syngas fuels tested in this study (Schwarze Pumpe, Exxon Singapore, and Tampa). The only syngas fuel that shows a 33% reduction of the mass of NO per unit energy generation is the PSI syngas fuel with high water volume fraction (22.6%) as shown in Table 1.

Baseline-CH$_4$

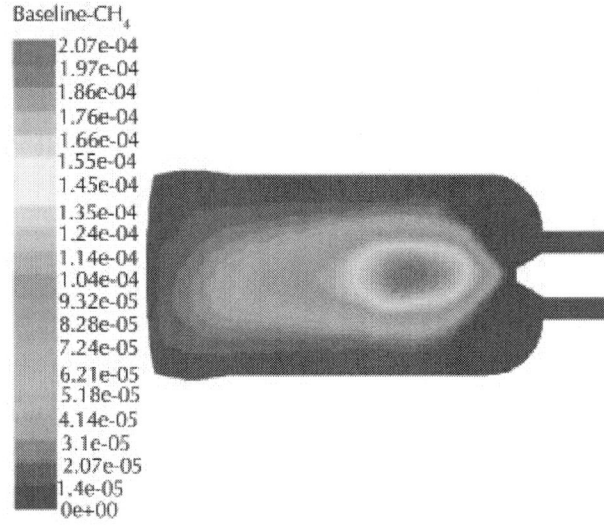

(a)

Syngas 1: Schwarze pumpe

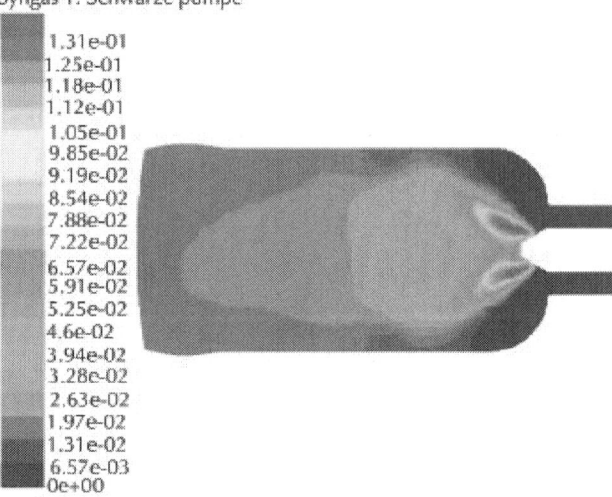

(b)

Syngas 5: Sarlux

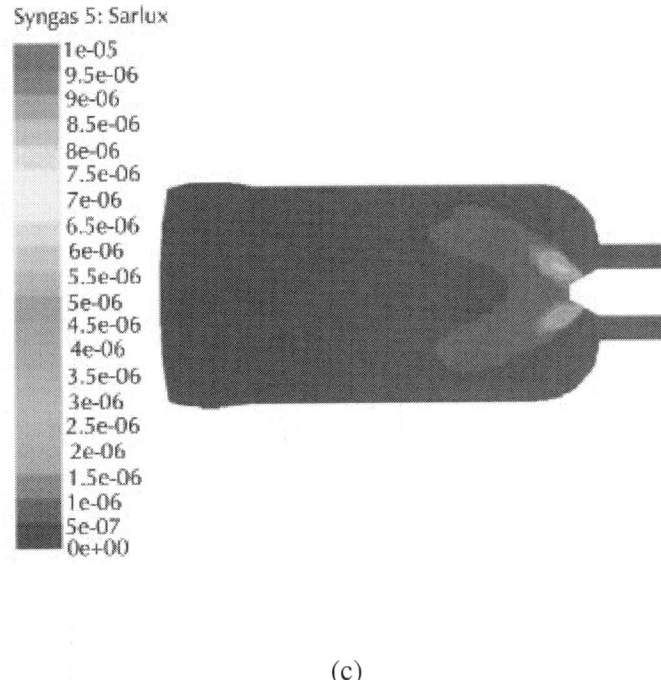

(c)

Figure 10: Contours of NO mass fraction.

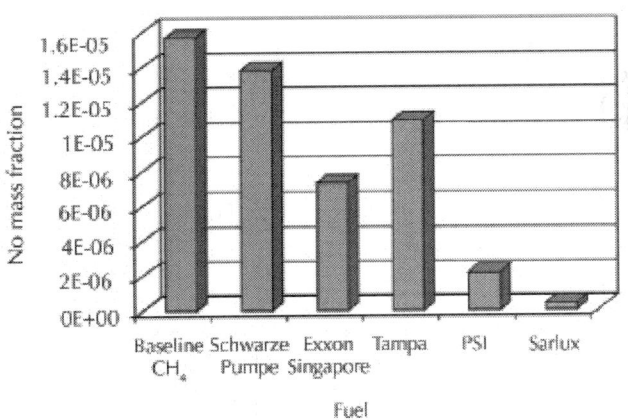

(a)

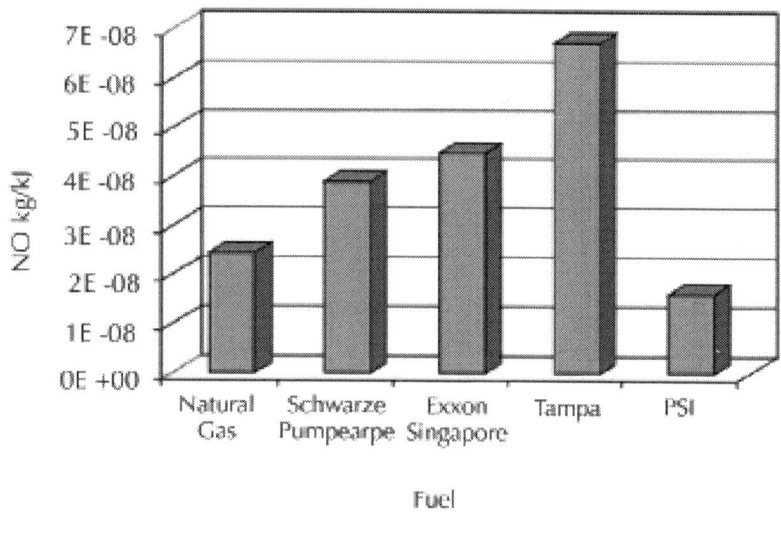

(b)

Figure 11: (a) Average NO mass fractions at the exit of the can combustor. (b) Average mass of NO at the exit per unit of energy generation.

It is noted that all the results presented in Figures 7 to 11 were obtained with a constant fuel flow rate. The syngas fuel composition and lower heating values were changed but the fuel mass flow rate was kept constant at 0.001 Kg/s. In fact when syngas heating value decreases, the mass fuel flow rate should increase to keep the same energy input in the can combustor. For the same fuel input in the can combustor, the fuel mass flow is four to five times greater than for methane or natural gas, due to the lower heating value of the syngas. Methane has a high lower heating value of 50.1 MJ/Kg. The lower heating values for the five syngas tested in this study is between 10.4 MJ/Kg and 27.8 MJ/Kg. Syngas is primarily composed of carbon monoxide and hydrogen but also contains a significant fraction (up to 50%) of non-combustibles (nitrogen, steam, carbon dioxide). The volumetric heating values of pure methane, hydrogen, and carbon monoxide are, respectively, 33.5, 10.2, and 12.6 MJ/m^3. The hydrogen and carbon monoxide has a lower volumetric heating value about

1/3 of the lower heating value of methane. When combined with nitrogen, water, and carbon dioxide in the gas stream, the syngas fuel volumetric heating value is even smaller (see Table 1). A comparison of the combustion process and emissions between the methane and syngas fuels with the same power and different fuel mass flow rate was investigated (see Table 3). The fuel mass flow rate for syngas 1 (Schwarze Pumpe) was increased from 0.001 Kg/s to 0.0018 Kg/s and syngas 3 (Tampa) was increased from 0.001 Kg/s to 0.0039 Kg/s to match the same heat input for the methane. It is noted that the heat input from methane combustion is 50.1 KW with a fuel mass flow rate of 0.001 Kg/s. The heat input from synags 1 (Schwarze Pumpe) with a fuel flow rate of 0.001 Kg/s and 0.0018 Kg/s is 27.8 KW and 50.1 KW. The heat input from synags 3 (Tampa) with a fuel flow rate of 0.001 Kg/s, and 0.0039 Kg/s is 12.7 KW and 50.1 KW. The fuel mass flow rates for syngas 1 and syngas 3 were increased, respectively, by a factor of 1.8 and 3.9 to match the same heat input for methane. The heat input (KW) was obtained by multiplying the lower heating value of the fuel (MJ/Kg) by the fuel mass flow rate (Kg/s). It is also noted that the primary and secondary air mass flow rates were increased accordingly to the increase in the fuel mass flow rate (see Table 3). Figure 12 shows the temperature contours of methane and syngas 3 with low and high fuel mass flow rate (same power of 50.1 KW). The results with the high fuel mass flow rate for syngas 3 or the same power show the same effects (lower gas temperature and shorter flame) as the results obtained with low fuel mass flow rate. The average masses of CO_2 and NO at the exit of the can combustor per unit of energy generation obtained with the same power were calculated, and the results are presented in Figures 13 and 14. With the same power, Figure 13 shows a reduction of the mass of CO_2 per unit energy generation of about 10.7% to 12.5% for Schwarze Pumpe syngas fuel compared to methane fuel but higher value of CO_2 for Tampa syngas fuel. Figure 14 shows a reduction of the mass of NO per unit energy generated for syngas fuel compared to methane fuel only when the primary and secondary air mass flow rates were increased according to the increase in the fuel mass flow rate (Schwarze Pumpe Syngas 1b and Tampa Syngas 3b) a shown in Table 3.

Table 3: Fuel and air injection conditions—same power generated

Constituents	Syngas 1a Schwarze pumpe	Syngas 1b Schwarze pumpe	Syngas 3a Tampa	Syngas 3b Tampa	Methane
Fuel mass flow rate (Kg/s)	0.0018	0.0018	0.0039	0.0039	0.001
Velocity of primary air (m/s)	10	18	10	39	10
Velocity of secondary air (m/s)	6	10.8	6	23.4	6
Total mass flow rate at the exit (Kg/s)	0.079	0.141	0.081	0.305	0.078
Lower Heating value MJ/Kg	27.8	27.8	12.7	12.7	50.1
Power KW	50.1	50.1	50.1	50.1	50.1

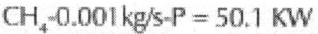

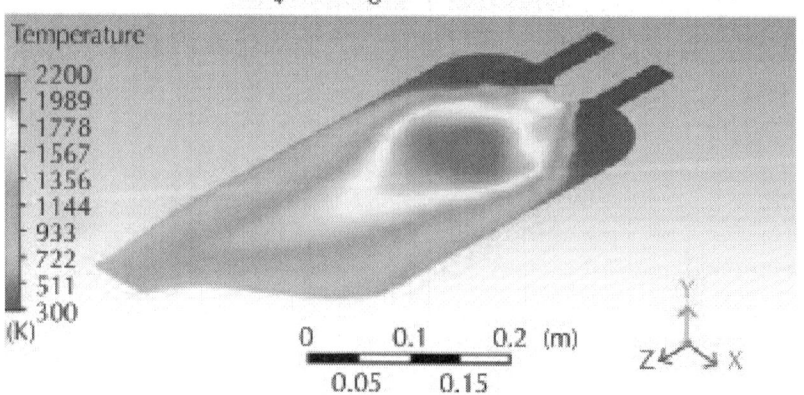

(a)

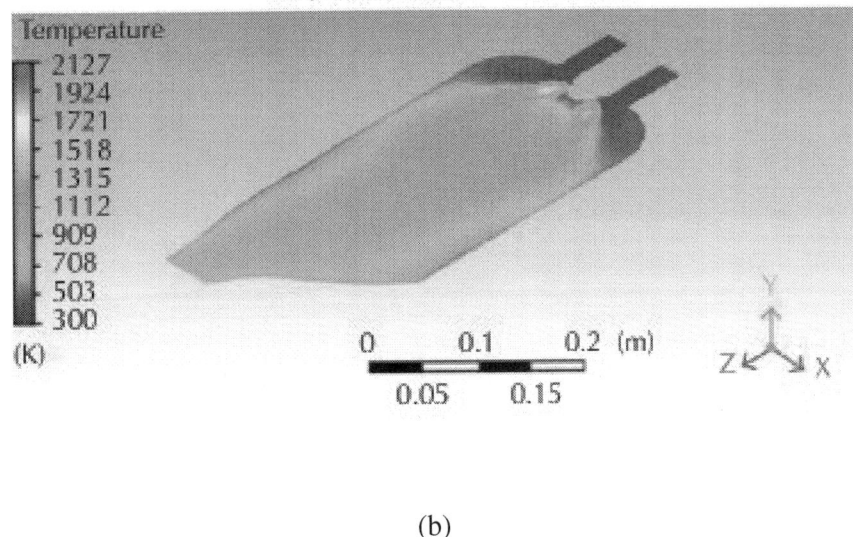

(b)

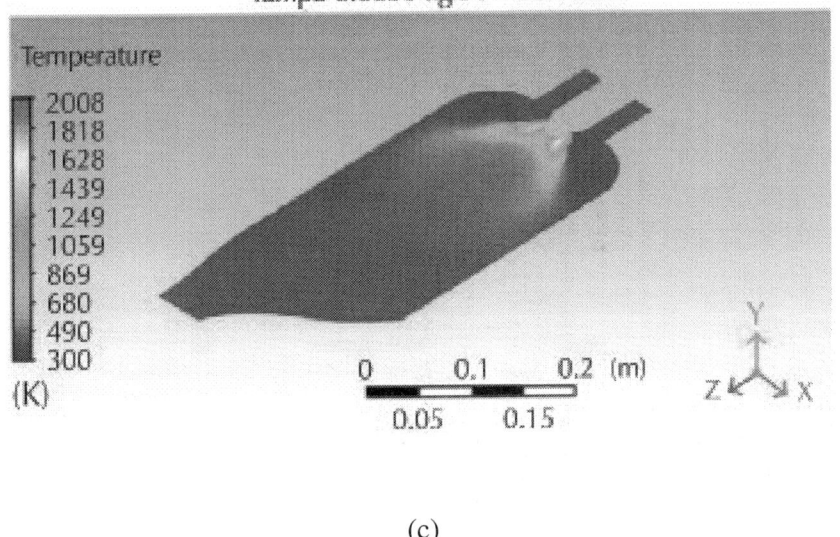

(c)

Figure 12: Syngas temperature contours: effect of fuel mass flow rate.

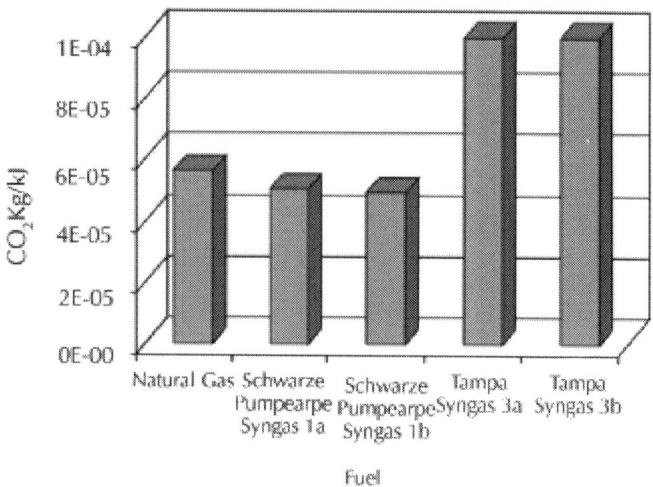

Figure 13: Average mass of carbon dioxide at the exit per unit of energy generation—same power 50.1 KW for all the fuels tested.

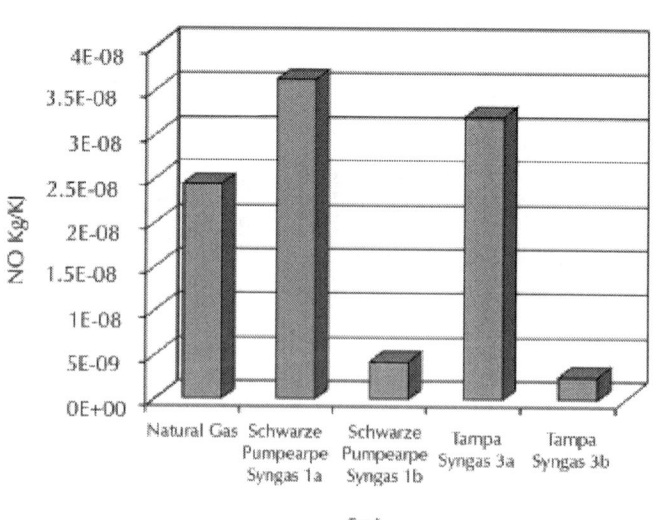

Figure 14: Average mass of NO at the exit per unit of energy generation—same power 50.1 KW for all the fuels tested.

CONCLUSIONS

Three-dimensional CFD analysis of syngas fuel combustion in gas turbine can combustor is presented in this study. Five Syngas fuel mixtures with different fuel compositions (H_2/CO = 0.63 to 2.63) and low heating values (8224 KJ/m^3 to 12492 KJ/m^3) were tested in this study. The syngas fuels are produced by different gasification processes using different feed stocks (coal, biomass, waste). The - model was used for turbulence modeling, mixture fractions/PDF model for non-premixed gas combustion and P-1 for radiation modeling. The effect of the syngas fuel composition (H_2, CO, CO_2, CH_4, N_2, and H2O) and fuel heating values on syngas flame shape, gas temperature, carbon dioxide (CO_2), and nitrogen oxides (NO_x) emissions was determined in this study

- Baseline fuel (methane) combustion: the results of the gas temperature, velocity field, swirling strength and CO_2 and NO_x emissions show that gas turbine can combustor burns the fuel efficiently, reduces the emissions, and, lower the wall temperature. The predicted maximum temperature of methane fuel combustion compares well with the theoretical adiabatic flame temperature.

- The gas temperature for the all five syngas shows a lower gas temperature compared to the temperature of methane. The gas temperature reduction depends on the lower heating value and the combustible constituents (hydrogen, carbon monoxide, and methane) and non-combustibles (inert) constituents in the syngas fuel.

- The results show a reduction (30% to 49%) of CO_2 mass fraction at the exit of the can combustor when methane is replaced with syngas fuel that produces less power (20% to 55% power reduction with syngas fuel). The reduction of CO_2 concentrations depends on the carbon monoxide and methane volume fractions in syngas fuel. For the same power generated by the methane and syngas fuels combustion, the results show a higher mass of CO_2 emitted per unit of energy generation (Kg/KJ) for Exxon Singapore, Tampa, and PSI syngas fuels compared to methane fuel. Schwarze Pumpe was the only syngas fuel that shows a reduction of about 12% of the average mass of CO_2 emitted per unit of energy generation (Kg/KJ) compared to methane fuel.

- With the same fuel mass flow rate but less power generated for the syngas fuel, the results show a reduction (11.5% to 97.6%) of the average NO mass fraction for synags fuel compared to methane. Syngas is used in non-premixed combustion to control the NO_x emissions by diluting the synags gas with nitrogen, carbon dioxide, and steam. The diluents reduce the flame temperature and consequently the formation of NO_x. For the same power generated by the methane and syngas fuels combustion, the results show a reduction of the mass of NO per unit energy generated for syngas fuel compared to methane fuel only when the primary and secondary air mass flow rates were increased according to the increase in the fuel mass flow rate to keep the same power.

The results obtained in this study show the change in gas turbine can combustor performance (temperature, flame shape, CO_2 emissions, and NO emissions) when natural gas or methane fuel is replaced by syngas fuels with lower heating value and fuel compositions.

REFERENCES

1. D. E. Giles, S. Som, and S. K. Aggarwal, "NOx emission characteristics of counterflow syngas diffusion flames with airstream dilution," Fuel, vol. 85, no. 12-13, pp. 1729–1742, 2006.

2. S. K. Alavandi and A. K. Agrawal, "Experimental study of combustion of hydrogen-syngas/methane fuel mixtures in a porous burner," International Journal of Hydrogen Energy, vol. 33, no. 4, pp. 1407–1415, 2008.

3. E. O. Oluyede, "Fundamental impact of firing syngas in gas turbines, gas turbine industrial fellowship program," Project Report, Electric Power Research Institute, Charlotte, NC, USA, 2006.

4. R. W. Schefer, "Combustion of hydrogen enriched methane in a lean premixed swirl burner," inProceedings of the DOE Hydrogen Program Review, 2001.

5.   S. Pater, Acoustics of turbulent non-premixed syngas combustion, Ph.D. thesis, University of Twente, Enschede, The Netherlands, 2007.

6.   S. Rahm, J. Goldmeer, M. Moilere, and A. Eranki, "Addressing gas turbine fuel flexibility," in Power-Gen Middle East conference, Manama, Bahrain, February 2009.

7.   R. D. Brdar and R. M. Jones, GE IGCC Technology and Experience with Advanced Gas Turbines, GE Power Systems, New York, NY, USA, 2000.

8.   P. Cheng, "Two-dimensional radiating gas flow by a moment method," AIAA Journal, vol. 2, pp. 1662–1664, 1964.

9.   R. Siegel and J. R. Howell, Thermal Radiation Heat Transfer, Hemisphere, Washington, DC, USA, 1992.

10. D. Todd and M. Gas, "Turbine Improvements enhance IGCC viability," in Gasification Technologies Conference, San Francisco, Calif, USA, October 2000.

Chapter 3

Physical Properties of Gas Hydrates: A Review

Jorge F. Gabitto[1] and Costas Tsouris[2]

[1]Department of Chemical Engineering, Prairie View A&M University, Prairie View, TX 77446-0397, USA

[2]Oak Ridge National Laboratory, Georgia Institute of Technology, Oak Ridge, TN 37831-6181, USA

ABSTRACT

Methane gas hydrates in sediments have been studied by several investigators as a possible future energy resource. Recent hydrate reserves have been estimated at approximately 10^{16} m^3 of methane gas worldwide at standard temperature and pressure conditions. In situ dissociation of natural gas hydrate is necessary in order to commercially exploit the resource from the natural-gas-hydrate-bearing sediment. The presence of gas hydrates in sediments dramatically alters some of the normal physical properties of the sediment. These changes can be detected by field measurements and by down-hole logs. An understanding of the physical properties of hydrate-bearing sediments is necessary for interpretation of geophysical data collected in field settings, borehole, and slope stability analyses; reservoir simulation; and production models. This work reviews information available in literature related to the physical properties of sediments containing gas hydrates. A brief review of the physical properties of bulk gas hydrates is included. Detection methods, morphology, and relevant physical properties of gas-hydrate-bearing sediments are also discussed.

INTRODUCTION

Clathrate hydrates or gas hydrates are solid structures. Water molecules are linked through hydrogen bonding and create cavities (host lattice) that can enclose a large variety of molecules (guests). No chemical bonding takes place between the host water molecules and the enclosed guest molecule. The clathrate hydrate crystal may exist at temperatures below as well as above the normal freezing point of water [1].

Clathrate hydrates of current interest are composed of water and the following molecules: methane, ethane, propane, isobutane, normal butane, nitrogen, carbon dioxide, and hydrogen sulfide. However, other nonpolar components between the sizes of argon (0.35 nm) and ethylcyclohexane (0.9 nm) can also form hydrates. Clathrate hydrates, commonly called gas hydrates, form at temperatures close to 273 K and elevated pressures [2].

The discovery of gas hydrates is credited to Sir Humphrey Davy [3] in 1810. Due to their crystalline, nonflowing nature, hydrates first became of interest to the hydrocarbon industry in 1934, when they were first observed [4] blocking pipelines. Hydrates concentrate hydrocarbons: 1 m³ of hydrates may contain as much as 180 SCM (standard cubic meters) of gas. Makogon [5] indicated that large natural reserves of hydrocarbons exist in hydrated form, both in deep oceans and in the permafrost. Evaluation of these reserves is highly uncertain, yet even the most conservative estimates concur that there is twice as much energy in hydrated form as in all other hydrocarbon sources combined [3].

Water (the host) molecules form the framework containing relatively large cavities as a result of hydrogen bonding; these cavities are occupied by gas (the guest) molecules, whose diameters are less than the size of the cavities. This hydrate structure is thermodynamically stabilized through nonbonded interactions between the encaged gas and the water lattice. Gas hydrates crystallize in three predominant structures: I (sI), II (sII), and H (sH), depending on the nature and the size of the guest molecule. There is a strong relationship between hydrate crystal structure and such properties as phase equilibria and heat of dissociation [2].

The following properties of sI and sII are determined by the molecular structure. (i) Mechanical properties approximate those of ice, perhaps because hydrates are 85 mol % water, yet each volume of hydrate may contain large volumes of the hydrate-forming species at standard temperature and pressure conditions (STP). (ii) Phase equilibrium is set by the size ratio of guest molecules within host cages, and three-phase (liquid water-hydrate-vapor; Lw-H-V) equilibrium pressure depends exponentially upon temperature. (iii) Heats of formation are set by the hydrogen-bonded crystals and are reasonably constant within a range of guest sizes [2].

Gas hydrate samples have currently been recovered from 19 or more areas worldwide and are believed to occur at about 77 locations including Antartica and Siberia (Kvenvolden and Lorenson [6]). In general, gas hydrates are found in marine shelf sediments and on-shore polar regions beneath the permafrost [7] because in these two types of settings the pressure-temperature conditions are within the hydrate stability field (Lerche and Bagirov [8]).

Offshore hydrate-bearing sediments have generally been found in waters deeper than 300 m; their zone of existence is from the seafloor to a depth of a few hundred meters, depending upon the local thermal gradient. Enormous amounts of methane are believed to be trapped by hydrates, both in the hydrate crystal structure itself and also in sediments beneath hydrate deposits [9].

The physical properties of gas hydrates trapped in sediments are very important in detecting the presence of these compounds, estimating the amount of gas hydrates trapped in the sediments, and developing processes to exploit this resource. Unfortunately, little is known about the physical properties of natural gas hydrate deposits in nature, making their detection by remote geophysical surveys difficult. The presence of gas hydrates in marine sediments dramatically alters some of the normal physical properties of the sediment, which can be detected by field measurements and by downhole logs [7].

The main goal of this work is to review some of the efforts made to determine physical property values for sediments partially, or completely, filled with gas hydrates. In order to achieve this goal, we will review the physical properties of bulk gas hydrates, the morphology of marine sediments, and models used to predict the properties of gas-hydrate-bearing sediments.

BULK GAS HYDRATES

Structures of Pure Gas Hydrates

The crystal structures of sI and sII hydrates were first determined in the late 1940s and early 1950s by von Stackelberg and coworkers using X-ray diffraction [3]. This discovery was followed by more extensive X-ray diffraction studies of these hydrate structures in 1965 by McMullan and Jeffrey [10] and by Mak and McMullan [11].

These structures differ in the number and sizes of the cages and in their unit cells (Table 1). The type of crystal structure that forms depends on the size of the guest molecule; for example, CH_4 and C_2H_6 both form sI hydrate, C_3H_8 forms sII hydrate, while larger guest molecules such as cyclopentane in the presence of methane form sH hydrate [11]. Both sI and sII hydrates have cubic crystal structures, while sH hydrate has a hexagonal crystal structure. All these hydrate structures are composed of two or more types of water cages packed within the crystal lattice. The water cages are described by the general notation X^n, where X = the number of sides of a cage face, and n = the number of cage faces having these X sides. It is not necessary for all cages to be occupied; for example, methane hydrate can be prepared with just 90% of the small cages occupied by methane. The sI and sII hydrate structures are of particular importance in the gas industry because they encage small gas molecules that are found in natural gas. The sI hydrate structure contains two different types of cavities: a pentagonal dodecahedral cavity (12-hedra), denoted 5^{12} (comprising 12 pentagons), and a larger tetracaidecahedral cavity (14-hedra), denoted $5^{12} 6^2$ (comprising 12 pentagons and 2 hexagons). The packing in sI hydrate can be described as 5^{12} cavities sharing vertices, with no direct face sharing occurring between these 12-hedra. The vertices of the 14-hedra are arranged in columns in which 12-hedra occupy the space between each pair of 14-hedra. The unit cell of sII hydrate also contains two different types of cavities: a small 5^{12} cavity and a hexacaidecahedral cavity (16-hedra), denoted $5^{12} 6^4$ (comprising 12 pentagons and 4 hexagons), which is slightly larger that the $5^{12} 6^2$ cavity found in sI hydrate. The packing in sII hydrate can be described as cavities sharing faces in 3D, with the void spaces being occupied by the 16-hedra (Koh [12]).

Table 1: Geometrical parameters of the main hydrate crystal structures (data taken from [2, 3, 12])

	Structure I		Structure II		Structure H		
Cavity	small	large	small	large	small	medium	large
Description	512	51262	512	51264	512	435663	51268
Cavities/unit cell	2	6	16	8	3	2	1
Average cavity radius, nm	0.395	0.433	0.391	0.473	0.391	0.406	0.571
Coordination number	20	24	20	28	20	20	36
Water moleculesper unit cell	46		136		34		
Lattice type	Cubic		Face-centered cubic		Hexagonal		
Unit cell parameters, nm	a=1.2		a=1.7		a=1.21, c=1.01		
Density, kg m-3	912		940		1952		

The hydrate structure sH, named for the hexagonal framework, was discovered [13] and shown by Ripmeester et al. [14] to have cavities large enough to contain molecules the size of common components of naphtha and gasoline. Physical properties, phase equilibrium data, and models have been advanced [15–18], and one instance [19] of in situ sH has been found in the Gulf of Mexico [12].

In 2001 a new hydrate structure, sT hydrate, has been discovered [20] in which all three types of large cages ($5^{12}6^3$, $5^{12}6^2$, and $4^15^{10}6^3$) in the structure are occupied by dimethyl ether guest molecules. The crystal structure in sT hydrate is trigonal [12]. In addition to the hydrate structures discussed above, two new structures have been discovered [21, 22] at higher pressures. Recently, Yang et al. (2009) [23] observed and characterized using X-ray diffraction and NMR a true clathrate hydrate containing only Xe atoms as guests. This form was produced from an initial pressurization of sI Xe hydrate followed by a temperature quench recovery. It is likely that more hydrate structure types will be discovered in the future. Table 1 provides a hydrate structure summary for the three most common hydrate unit crystals (sI, sII, and sH).

Physical Properties of Bulk Gas Hydrates

Sloan [2] provided a review of time-independent physical/chemical properties as they relate to crystal structures. According to Sloan, if all the cages of each structure are filled, all three structures (sI, sII, and sH) of hydrates have the property of being approximately 85% (mol) water and 15% gas. The fact that the water content is so high suggests that the mechanical properties of the three hydrate structures should be similar to those of ice. This conclusion is true to a first approximation, as shown in Table 2, with the exception of thermal conductivity and thermal expansivity [24, 25]. Sloan [2] proposed that the guest/cavity size ratio provides an adequate basis to understand the physical properties of bulk gas hydrates. The concept of "a ball fitting within a ball" was proposed as a first approximation. Sloan enunciated the following five points regarding the guest-cavity size ratio for hydrates formed from a single guest component in sI or sII.

- The size of the stabilizing guest molecules ranges between 0.35 and 0.75 nm. Molecules smaller than 0.35 nm will not stabilize the sI structure, and molecules larger than 0.75 nm, will not stabilize the sII structure.
- Some molecules are too large to fit the smaller cavities of each structure.

- Other molecules such as CH_4 and N_2 are small enough to enter both the small and large cavities when hydrate is formed from these single components.
- The larger molecules of a gas mixture usually determine the structure that is formed. The larger components enter into large cavities only, while smaller components enter both cavities.
- Molecules which are very close to the lines separating the cavity sizes change the stoichiometry, due to their inability to fit comfortably within the cavity.

Table 2: Comparison of properties of ice, sI, and sII hydrate crystal structures (data taken from [2, 3])

Property	Ice (Ih)	Structure I	Structure II
Water molecules number	4	46	136
Lattice parameters at 273 K, nm	a=0.452, c=0.736	1.20	1.73
Dielectric constant at 273 K	94	~58	58
Water diffusion correlation time, μs	220	240	25
Water diffusion activation energy, kJ/m	58.1	50	50
Isothermal Young's modulus at 268 K, 109 Pa	9.5	8.4 (est.)	8.2 (est.)
Poisson's ratio	0.33	~0.33	~0.33
Bulk modulus (272 K)	8.8	5.6	NA
Shear modulus (272 K)	3.9	2.4	NA
Compressional velocity (Vp), m/s	3870.1	3778.0	3821.8
Shear velocity (Vs), m/s	1949	1963.6	2001.1

Velocity ratio (comp./shear)	1.99	1.92	1.91
Linear thermal expn., at 200 K, K-1	56 10-6	77 10-6	52 10-6
Adiab. bulk compress. (273 K), 10-11 Pa	12	14 (est.)	14 (est.)
Heat Capacity, J kg-1 K-1	3800	3300	3600
Thermal conductivity (263 K), W m-1 K-1	2.23	0.49±0.02	0.51±0.02
Refractive index, 638 nm, -3°C	1.3082	1.3460	1.350
Density, kg m-3	916	912	940

Sloan [2] noted that a size ratio (guest molecule size divided by the cavity size) of approximately 0.9 is necessary to achieve stability of a hydrate structure. When the size ratio exceeds unity, the molecule will not fit within the cavity and the structure will not form. When the ratio is less than 0.9, the molecule cannot lend significant stability to the cavity [2].

Phase equilibrium is set by the size ratio of guest molecules within host cages, and the three-phase (Lw-H-V) equilibrium pressure depends exponentially upon temperature. Heats of formation are set by the hydrogen-bonded crystals and are reasonably constant within a range of guest sizes [2].

Thermal conductivities of a few gas hydrates have been published by Cook and Leaist [26] (1981), Ross et al. [27] (1978), Stoll and Bryan [28] (1979), and Ashworth et al. [29] (1985). These values are much smaller than those of ice (Ih) both at and well bellow the freezing point (Tse and White [30], 1988).

In contrast with most well-defined crystalline structures, in which the thermal conductivity falls with increasing temperature following a T^{-1} dependence (for T>100 K), the thermal conductivity of clathrate hydrates increases slightly with increasing temperature (Tse and White [30]). The thermal conductivity of clathrate hydrates is five times lower than that of ice near the melting point, and even lower (by a factor >20) at lower temperatures. The temperature dependence of thermal conductivity in clathrate hydrates is characteristic of an amorphous material (Tse and White [30]).

Based on the low-temperature thermal conductivity of tetrahydrofuran (THF) clathrate hydrates, Anderson and Suga [31] suggested that, despite the well-defined crystalline structures, clathrate hydrates show glassy behavior attributable to low-frequency guest vibrations, causing the clathrate hydrates to be thermal glasses.

From measurements of thermal conductivity at relatively high temperatures, it appears that the unusual thermal conductivity is insensitive to the crystal structure and dependent on the host. As a result of all the measurements of thermal conductivity of clathrate hydrates, it can be concluded that these compounds are very poor thermal conductors. Furthermore, the large difference in the thermal conductivity of pure ice and clathrate hydrates provides a criterion for locating regions of shelf ice which contain potential energy reserves in the form of methane or similar gases. Thermal conductivity is also a vital parameter required for modeling the recovery of natural gas from hydrates; knowledge of the variation of the thermal conductivity with pressure and temperature is required for successful methane hydrates exploitation [32].

The heat of dissociation (ΔH_d) is defined as the enthalpy change to dissociate the hydrate phase to a vapor and aqueous liquid, with values given at temperatures just above the ice point. For sI and sII, to a fair engineering approximation of 10%, it has been shown [3] that ΔHd is a function of the number of crystal hydrogen bonds (loosely taken as hydration number). However, the value of ΔH_d is relatively constant for molecules which occupy the same cavity, within a wide range of component sizes. Enthalpies of dissociation may be determined via the univariant slopes of phase equilibrium lines (ln P versus 1/T) in the previous paragraphs, using the Clausius-Clapeyron relation [$(\Delta H_d) = $ -zRD (ln P)/d1/T] (Sloan [2]).

GAS-HYDRATE-BEARING SEDIMENTS

Seismic Detection Methods

The presence of gas hydrates in offshore continental margins has been inferred mainly from seismic processing techniques. Seismic image processing visualizes the subsurface structure by means of reflected acoustic signals. The seafloor signal is marked by a white/black reflection, which means that the subsurface volume is harder than the volume above. In acoustic terms, the acoustic impedance (the product of medium density and speed of sound) below the seafloor in the sediment is higher than the impedance of the water column. In contrast, the bottom-simulating reflector (BSR) is marked by a black/white reflection indicating possibly high hydrate impedance above gas-filled sediments with low impedance. As the resolution of the seismic image is limited by the seismic source bandwidth and as the physical parameters describing the seismic subsurface response are frequency dependent, multiple surveys with different acquisition parameters are needed to obtain a more complete knowledge of the sediment parameters [33].

Seismic measuring arrays are composed of a sound source, such as air guns (usually a dozen or more of them), and moving or stationary receivers; the reflected sound waves provide a 2D image of a slice through the earth's surface. The sound waves travel through the water column and back as compression or P waves, representing the vertical motion through the different materials [34]. There is also interest in other seismic methods, such as the use of shear or S waves. Shear waves have different vector components of horizontal movement and are converted from compression waves passing through gas hydrate zones. Because compression and shear waves deform materials differently, some researchers are using these differences to learn more about the amount and distribution of gas hydrate [35].

Locating likely areas of gas hydrates using remote seismic sensing is relatively straightforward in many parts of the world where the BSR is readily evident. Because the BSR follows a thermobaric surface rather than a structural or stratigraphic interface, it is normally observed to crosscut other reflectors. The BSR usually marks the base of a gas hydrate layer below which there is free gas. However, hydrate samples have been taken in areas where no BSR was found. One reason is that because the BSR represents the base of stable hydrates and is the seismic image of the interface between solid hydrate and free gas, where no free gas is present, there is no BSR even though hydrates may well be present [36]. Furthermore, it is difficult to quantify the volume fraction of hydrates in the shallow section from seismic data alone. Gas hydrates are electrically more resistive than the host sediments; therefore, they will have an electromagnetic signature that increases with hydrate volume fraction. Indeed, well logs indicate increased resistivity in zones of gas hydrate, although this effect is sometimes modest. Marine electromagnetic (EM) methods have long been proposed as an effective way to map and characterize gas hydrates, notably by Nigel Edwards at the University of Toronto. The Hydrate Ridge work represented the first attempt to apply the EM methods recently developed for oil field characterization to the hydrate question. The data are promising, but this survey was a pilot study with less than 4 days of ship time available on station. Furthermore, the physical characteristics of in situ hydrate vary considerably, and Hydrate Ridge, while a good test of the method because of the extensive seismic and drilling data sets available, may not be characteristic of hydrates in more commercially relevant areas, such as the Gulf of Mexico [37].

Hydrate Morphology in Sediments

Gas hydrate morphology describes the relationship between gas hydrates and the surrounding marine sediments. The morphology of gas hydrates determines the basic physical properties of the sediment-hydrate matrix. Many remote techniques for gas hydrate detection and quantification are highly dependent on the hydrate morphology. Little attention has been paid to the hydrate morphology until now because previous methods of hydrate collection preserved only the grossest morphologies (e.g., lumps and nodules of hydrate [38]). However, recent advances in pressure coring and pressure core analysis have allowed collection of samples with intact gas-hydrate-sediment morphologies, which show that gas hydrates often take on complex forms which will require new approaches to both conceptualizing and modeling gas hydrate dynamics. The recent use of borehole pressure coring tools has allowed marine gas-hydrate-bearing sediments to be recovered containing centimeter to sub-millimeter gas hydrate structures preserved in their in situ condition [38].

Clear knowledge of detailed gas hydrate morphology will provide critical data to help determine the mechanisms of gas hydrate deposit formation and to model the kinetics of deposit dissociation from both natural and artificial causes. The morphology also has a significant effect on sedimentary physical properties, from seismic velocities on a large scale to borehole electrical resistivity on a smaller scale, and will thus impact the amount of gas hydrate saturation estimated from geophysical data [38].

Malone (1985) [39] described four possible hydrate morphologies using the terminology disseminated, nodular, vein, and massive. Disseminated hydrate occurs within the pore space of the sediment, while the other three forms occur where the sediment is disturbed either by regional tectonic stresses or through the stress resulting from hydrate crystal growth. For instance, Cook and Goldberg (2007) [40] found hydrate-bearing fractures to be oriented with respect to regional tectonic stresses offshore India. However, some researchers have not found this classification useful to further our understanding of sediment–hydrate interactions [41].

Theoretical work on hydrate formation (Clennell et al. [42]; Henry et al. [43]) explored the influence of capillary pressure and thermodynamics on hydrate growth and provided some real physical constraints to hydrate morphology. It was concluded that hydrate growth in fine-grained mud would be unlikely, and that coarser-grained sediments, exhibiting larger pores, would act as more likely hosts. Theoretical considerations suggested that hydrate morphology is controlled by the nature of the sediment host as much as by the supply of the necessary chemical components and environmental conditions (water, gas, nucleation sites, temperature, and pressure). This theoretical work is also supported by experimental studies (Kleinberg et al. [44], 2003; Camps [45], 2007) and by observations (e.g., Riedel et al. [46], 2006) where disseminated hydrate is limited to coarser-grained sediments and the other forms tend to occur in finer-grained sediments where the sediment fabric is disturbed. Similar observations were found through investigations of hydrate dissociation conditions (Anderson et al. [47], 2003; Llamedo et al. [48], 2004). These observations revealed that dissociation is more readily achieved within small pores as compared to large pores, suggesting the possibility of hydrate breakdown in small pores rather than in adjacent large pores [41].

Significant advances in characterizing the relationship between sediments and hydrates have been recently achieved [38]. For instance, Tinivella et al. [49] have quantified the concentrations of gas hydrate in pore space by travel-time inversion modeling of the acoustic properties of these sediments. Such analysis has allowed the identification of free gas distribution in pore spaces, likely patterns of fluid migration, the physical properties of sediments, and the consequent origin of the BSR offshore the Antarctic Peninsula [41].

At other sites, hydrate volumes have often been estimated using electrical or acoustic measurements and relating these parameters (electrical resistivity and acoustic velocity) to the hydrate concentration or saturation in the pore space. Such studies usually assume the hydrate is disseminated in the pore space and that the sediment remains wet, comparable to oil-bearing reservoirs. Ecker et al. [50] and Dvorkin et al. [51] demonstrate that knowledge of the interaction between hydrate and sediment grains is crucial in achieving well-constrained volume estimates of hydrate. It is also recognized that the use of Archie's equations [52] assumes the hydrate does not completely block off the pore space at low saturations, treating hydrate as a hydrocarbon fluid [41].

The estimation of the hydrate content using down-hole electrical measurements based on Archie's law requires the knowledge of the saturation exponent. The saturation exponent is an empirical parameter that includes influences from the internal rock structure such as pore shape, size, connectivity of the pore network, and the distribution of the conducting phase. Spangenberg [53] used different models, which account for different morphological forms of gas hydrates, to study the influence of gas hydrate content on the electrical properties of the hydrate-bearing sediment. The author concluded that for all studied forms of hydrate occurrence, disseminated in the pore space, nodular, and layered, the saturation exponent depends on the sediment properties and on saturation itself. The growth of gas hydrate nodules, lenses, and layers is a process that is assumed to result in the displacement and compaction of the surrounding sediment. Because of this change of sediment properties during hydrate generation, the saturation exponent for these forms of hydrate occurrence depends strongly on the relationship between porosity and formation resistivity factor, expressed in the form of Archie's cementation exponent. For the case that hydrate occurs disseminated in the pore space and the assumption that capillary effects are important for hydrate generation, the saturation exponent depends on grain size and grain size sorting. Spangenberg [53] reported that, for the parameters used in his model calculations, the saturation exponent varies between 0.5 and 4.

Holland et al. [38] proposed that gas hydrate morphologies are found in two basic types: pore filling and grain displacing. Pore-filling morphologies of gas hydrate replace pore fluid between grains of sediment; this gas hydrate may or may not cement grains together. Grain-displacing gas hydrate does not occupy the pore volume between grains but, instead, forces grains apart, forming veins, layers, and lenses of pure gas hydrate. Grain-displacing hydrate may cover a vast range of sizes, from thin veins of possibly only a few microns thick to nodules of tens of centimeters or even meters in diameter. Holland et al. suggested that "grain displacing" and "pore filling" are not equivalent to the terms "massive" and "disseminated" but that these terms apply to cores which have already undergone gas hydrate dissociation, where massive gas hydrate is still visible, and disseminated gas hydrate is invisible to the naked eye, and may have already been completely dissociated [38].

Tohidi et al. [54] conducted visualization experiments using two-dimensional transparent glass micro-models and reported that hydrates can form from either free or dissolved gas. They also reported that hydrates usually form within the center of pore spaces, with a thin film of water covering the grains, rather than nucleating on grain surfaces.

Kingston et al. [55] used a specially constructed laboratory porous medium, gas hydrate resonant column (GHRC), to explore different methods of hydrate synthesis and measure the properties of the resulting sediments. The authors studied different water saturation conditions of the porous medium. In low water saturation tests, or conditions where the environment is gas saturated, the hydrate will grow on the water location; therefore, the water saturation becomes the restricting factor on hydrate content. In partially saturated sands, water tends to collect at grain contacts and coat individual sand grains. Hydrate will therefore preferentially grow at grain contacts, effectively cementing the rock grains. As hydrate content is increased, it begins to fill the pores, but the increased stiffening is likely to be produced by the increased quantity of "cement" at grain contacts. In fully water saturated tests, the morphology of the hydrate appears to be different. Under these conditions, the presence of gas bubbles suspended in water-filled pores was reported before hydrate formation. In this case, hydrate will form at the gas/water interface (i.e., around gas bubbles). The hydrate has now become a pore-filling component, and only large amounts of hydrate in the pore space will have a significant effect on the physical properties [55].

The production of natural gas from oceanic and permafrost sediments is currently being developed using such methods as depressurization, thermal stimulation, and injection of hydrate inhibitors (Moridis et al. [56], 2004). It is important to understand the physical properties of sediments in investigations of structural properties, such as permeability, hydrate saturation, and sediment porosity, since these properties are essential to the development of natural gas production. The porosity is particularly important for material flow in sediment due to the relationship between porosity and permeability (Koponen et al. [57]; Singh and Mohanty [58], Bernabè et al. [59]; Arns et al. [60], among others). Hydrate saturation is required to estimate the physical properties of sediments.

Different experimental techniques have been used to determine these structural parameters. Jin et al. [61] used microfocus X-ray computed tomography (CT) to study the structure of natural gas sediments with and without gas hydrates. The authors applied a newly developed high-pressure vessel for the microfocus X-ray CT system to observe the sediments at a temperature above 273 K and under high-pressure conditions. Jin et al. [62] used two-dimensional CT images to measure the spatial distribution of the free-gas, sand particles, liquid water, and solid hydrate phases.

Jin et al. [62] assessed the permeability of sediments via the correlation between the absolute permeability and the pore network in sediments. The continuous pore channel, which allows gas and water flow, was analyzed from the three-dimensional sediment images using a microfocus X-ray computed-tomography system. Their results showed that the proportion of the horizontal-continuous pore channel in terms of direction is a dominant factor in determining the absolute permeability. The absolute permeability of the sediment correlated well with the distribution of the continuous pore channel.

Minagawa et al. [63] used proton nuclear magnetic resonance (NMR) measurements combined with a permeability measurement system to characterize methane-hydrate-bearing sediment based on pore-size distribution and permeability. They compared the effective permeability of sediments with different effective porosities, which had been measured by water flow based on Darcy's law, with the permeability calculated by NMR spectra based on the SDR (Schlumberger-Doll Research) model. The permeability calculated by both methods was similar, with a difference between them of less than a factor of 2. Minagawa et al. [63] used their results to describe the relation between pore-size distribution, porosity, and effective permeability.

The presence of pore-scale phenomena, however, could introduce additional complexities. In particular, the effect of pore space geometry; pore-network topology and heterogeneity; and pore space accessibility on gas-phase mobility can influence, among other issues, the ability to economically produce methane gas. To further emphasize the importance of porous sediments, consider that typical pore values found in oceanic sediments are very low. For example, Yang and Alpin [64] reported for mudstones obtained from the Norwegian margin mean pore-throat sizes in the range 8–452 nm and pore-throat size standard deviation in the range 9–1425 nm. However, hydrate occurrence, in addition to fine-grained sediments, has also been reported in coarser oceanic sediments. A more detailed discussion of this issue is given by Tsimpanogiannis and Lichtner [65], who also considered larger pore values corresponding to onshore sediments under the permafrost as well.

Physical Properties of Gas-Hydrate-Bearing Sediments

The presence of gas hydrates in marine sediments dramatically alters some of the normal physical properties of the sediment, which can be detected by field measurements and by down-hole logs [3]. An understanding of the physical properties of hydrate-bearing sediments is necessary for interpretation of geophysical data collected in field settings, borehole, and slope stability analyses; reservoir simulation; and production models.

The physical properties of bulk hydrates are remarkably close to those of pure ice: the compression and shear (P and S) wave velocities in methane hydrate may reach 3600 and 1900 m/s, respectively, while its density is 0.912 g/cc. The corresponding values for ice are 3890 and 1970 m/s and 0.916 g/cc, respectively. As a result, the properties of sediments containing hydrate in the pore space are similar to sediments containing normal ice. However, these sediments are much more rigid than sediments filled solely by water, and unlike ice, methane hydrate can be ignited. A unit volume of hydrate releases about 160 unit volumes of methane (under normal conditions). Also, unlike ice, hydrate can exist at temperatures above 32°F (0°C), but it requires high pore pressure to form and remain stable [66].

Hydrates normally exclude the salt in the pore fluid from which it forms, and thus they have high electric resistivity just as ice and sediments containing hydrates have a higher resistivity compared to sediments without gas hydrates (Judge [67]). The unconsolidated sediments in the upper several hundred meters of the marine sediment section (50% porosity) normally have a very low resistivity of about 1 ohm-m. For 15–20% hydrate saturation in the pore space (7–10% of sediment), the resistivity increases by about a factor of 2. The most readily observable change in sediment physical properties resulting from the formation of gas hydrates is an increase in seismic velocity. Laboratory-measured seismic velocities for porous media at the maximum hydrate saturation vary from 2700 to 6000 m/s, depending on the type of sediment and the method of preparation (Stoll et al. [68]; Stoll [69]; Pearson et al. [70]), compared to about 3600 m/s in pure methane hydrate [71]. These laboratory determinations are for sediments of much lower porosity and higher velocity than most continental margin sediments near the seafloor. Sediments containing substantial hydrates have enhanced velocities. For example, a hydrate saturation of 10–20% of pore space in unconsolidated sediment (50% porosity at depths of a few hundred meters) has a velocity of 1900–2100 m/s, compared to no-hydrate velocities of 1600–1700 m/s. In general, if hydrates occupy 15% of pore space, a 15–20% increase in sediment seismic velocity is expected. This increase can be detected in interval velocities from high-quality multichannel seismic data and in well-calibrated down-hole sonic logs. A small quantity, 1-2%, of free gas in the sediment pore space beneath the BSR will significantly reduce the sediment velocity, while a further increase in gas concentration makes little change (Murphy [72]). The effect of free gas on sediment velocity is highly dependent on water depth primarily because gas density and compressibility are very sensitive to pressure and temperature [36].

Numerous publications report on laboratory measurements of the geophysical and geotechnical properties of hydrate-bearing sediments, but many fundamental challenges remain in using this information to interpret borehole logs or other field data obtained in hydrate zones. The difficulty of maintaining hydrate-bearing sediments within the hydrate stability field has led some researchers to construct specialized devices for their experiments that can be used to reproduce results in other laboratories under exactly the same set of experimental conditions. Santamarina et al. [73] undertook an exhaustive series of laboratory measurements to determine the mechanical, thermal, electrical, and electromagnetic properties of hydrate-bearing soils using standardized geotechnical devices and test protocols. They conducted experiments on soils with a range of grain sizes subject to an effective stress of up to 2 MPa and with well-controlled saturations of synthetic hydrate [74].

Stoll and Bryan [28] carried out an experimental study to determine the thermal conductivity and acoustic wave velocity in hydrates and sediments containing hydrate. The most significant result of their studies was that the formation of hydrate decreased the thermal conductivity in sediments. This behavior is contrary to what might be expected when compared with the behavior of frozen sediments. Also, based on measurements of acoustic wave velocity, this research confirmed that both pure water and water-bearing sediment are converted to a stiff elastic mass by the formation of a sufficient amount of hydrate. Stoll and Bryan concluded that this finding served as the basis for using the sharp acoustic impedance contrast at the boundary of a hydrate-containing sediment to locate hydrate deposits [28].

Pearson et al. [71] predicted the physical properties of hydrate-containing sediments in order to include their effects on production models and to develop geophysical exploration and reservoir characterization techniques. The authors used empirical relationships between ice composition and seismic velocity, electrical resistivity, density, and heat capacity developed for frozen soils to estimate the physical properties of hydrate deposits. They proposed that the resistivity of laboratory permafrost samples follows a variation of Archie's equation [52]:

$$\frac{\rho_f}{\rho_t} = C^{-T} S_w^{1-n},$$

(1)

where ρ_f and ρ_t are the thawed and frozen resistivities of the sample, T is temperature, S_w is the unfrozen water content, and n and C are empirical constants. The parameters C and N were calculated for a variety of lithologic types.

Pearson et al. [71] also reported that the compressional wave velocities of partially frozen sediments (V_p) are related to the velocities of the matrix (V_m), the liquid (V_l), and solid (V_s) phases present in the pores by the well-known three-phase rule:

$$\frac{1}{V_p} = \left[\varepsilon \frac{S_w}{V_L}\right] + \left[\varepsilon \frac{(1 - S_w)}{V_s} + \frac{(1 - \varepsilon)}{V_m}\right],$$

(2)

where ε is the porosity.

Pearson et al. proposed that the resistivities and seismic velocities are both functions of the unfrozen water content; however, it was found that resistivities are more sensitive to changes in S_w and can vary by as much as three orders of magnitude, which allows the use of electrical resistivity measurements to estimate the amount of hydrate in place. Pearson et al. [71] estimated the unfrozen water content, assuming that the dissolved salt in the pore water concentrates as a brine phase as the hydrate forms. Using this technique, they estimated the brine content as a function of depth, assuming a wide range of temperature gradient and pore water salinity values. They also reported that the presence of hydrates tends to lower the heat capacities and densities of sediments, even though these effects are comparatively small.

Pearson et al. [70] reported laboratory acoustic velocity and electrical resistivity measurements on Berea Sandstone and Austin Chalk samples saturated with a stoichiometric mixture of tetrahydrofuran (THF) and water. THF solutions were selected because they form hydrates similar to natural gas hydrates at atmospheric pressures. The authors reported that hydrate formation in both the chalk and sandstone samples increased the acoustic P-wave velocities by more than 80% when hydrates formed in the pore space; however, the velocities soon became constant, and further lowering the temperature did not appreciably increase the velocity. In contrast, the electrical resistivity increased nearly two orders of magnitude upon hydrate formation and continued to increase slowly as the temperature was decreased. The dielectric loss showed a linear decrease with frequency suggesting that ionic conduction through a brine phase dominates at all frequencies, even when the pores are nearly filled with hydrate. The authors also reported that resistivity values were a strong function of the dissolved salt content of the pore water. Pore water salinity also influenced the sonic velocity, but this effect was much smaller.

Winters et al. [75] measured a wide range of acoustic P-wave velocities (V_p) in coarse-grained sediments for different pore space occupants. The measured V_p values ranged from less than 1000 m/s for gas-charged sediments to 1770–1940 m/s for water-saturated sediments, 2910–4000 m/s for sediments with varying degrees of hydrate saturation, and 3880–4330 m/s for frozen sediment. V_p values measured in fine-grained sediments containing gas hydrate were substantially lower (1970 m/s). The presence of gas hydrate and other solid pore-filling material, such as ice, increased the sediment shear strength. The magnitude of that increase is related to the amount of hydrate in the pore space and cementation characteristics between the hydrate and sediment grains. Winters et al. found that, for consolidation stresses associated with the upper several hundred meters of sub-bottom depth, pore pressures decreased during shear in coarse-grained sediment containing gas hydrate, whereas pore pressure in fine-grained sediment typically increased during shear. The presence of free gas in pore spaces damped pore pressure response during shear and reduced the strengthening effect of gas hydrate in sands [75].

The research group led by Santamarina collected experimental data to conduct a comprehensive analysis of the values of geophysical and geotechnical properties as a function of hydrate saturation, soil characteristics, and other parameters. Measurements of seismic velocities, electrical conductivity and permittivity, large strain deformation and strength, and thermal conductivity were emphasized in these experiments (Fernandez and Santamarina [76]; Santamarina et al. [74]; Yun et al. [77]; Santamarina et al. [73]).

Santamarina et al. [73] used their data set to identify the systematic effects of sediment characteristics, hydrate concentration, and state of stress, developing mathematical relations for the most relevant material parameters. They reported that, under low strain conditions, the shear wave velocity in hydrate-bearing sediments is stress dependent at low hydrate concentrations but becomes hydrate controlled at high hydrate concentrations. The P-wave velocity (V_p) in hydrate-bearing sediments can be computed from the shear wave velocity of the hydrate-bearing sediment (V_{hbs}) and the volume fraction and bulk stiffness (B_j) of the component phases. Following a Biot-Gassmann-type formulation for low skeletal stiffness ($B_{sk}/B_w \ll 1$), the following relationship can be used (Santamarina et al. [78]):

$$V_p^2 = V_{hbs}^2 \left(\frac{1 - \nu_{sk}}{1 - 2\nu_{sk}} + \frac{4}{3} \right)$$
$$+ \frac{1}{\rho_{mix}} \left(\frac{1 - \varepsilon}{B_{mix}} + \varepsilon \left(\frac{S_h}{B_h} + \frac{1 - S_h}{B_w} \right) \right)^{-1}.$$

$$(3)$$

The researchers noted that the small strain Poisson's ratio for the skeleton V_{sk} is typically 0.1 ± 0.05. The shear wave velocity in hydrate-bearing sediments (V_{hsb}) is stress dependent at low hydrate concentrations but becomes hydrate controlled at high hydrate concentrations. Following these data-based observations, and adopting the form of theoretical expressions for cemented soils (Fernandez and Santamarina [76]), Santamarina et al. [78] fit the data using the following relationship:

$$V_{hbs} = \sqrt{ \left(\frac{V_h S_h^2}{\varepsilon} \right)^2 \theta + \left[\alpha \left(\frac{\sigma'_{||} + \sigma'_{\perp}}{2\ \text{kPa}} \right)^\beta \right]^2 }.$$

$$(4)$$

Here, $\sigma'_\perp$ and $\sigma'_\parallel$ are the effective stresses acting in the direction of wave propagation ($\parallel$) and particle motion ($\perp$), respectively; the factor $\varnothing$ represents the hydrate influence in the pore space; and   and   are factors calculated from tests conducted on sediment without hydrates ($S_h=0$).

Under high strain conditions, it was found that (1) the undrained shear strength (Su) at low hydrate concentration is determined by the effective stress-dependent frictional strength, (2) the contribution of the hydrate strength increases nonlinearly with higher strength, gaining relevance at high S_h, and (3) in the case of fine-grained soils, the effect of hydrate tends to be more pronounced at low porosity. The following expression for S_u captures these observations:

$$S_u = a\,\sigma'_o + bq_h \left(\frac{S_h}{\varepsilon}\right)^2.$$

(5)

Here a nominal value for the hydrate strength, $q_h=8$ MPa, was assumed. This value is within the range reported in the literature [79]. The coefficient a represents friction and pore pressure generation in the sediment, while b is an indication of the hydrate's ability to contribute to the strength of the hydrate-bearing sediment.

The electrical conductivity of hydrate-bearing sediments, σ_{hsb}, at radiofrequencies is determined by the volume fraction of the unfrozen pore fluid and the pore fluid conductivity (σ_w). The resulting expression fitted their experimental data:

$$\sigma_{hbs} = \sigma_w \left[\varepsilon\,(1 - S_w)\right]^{1.6}.$$

(6)

The electrical permittivity (k_{hsb}) in the microwave frequency range is determined by the polarization of the free, unfrozen water. Most of their experimental data at $S_w=0.5$ were fitted by

$$\kappa_{hbs} = 5 + 70 \; \varepsilon \; (1 - S_h).$$

(7)

The thermal conductivity (k_{hsb}) was determined using the needle probe technique in sediments subjected to isotropic confinement [32]. The experimental data showed that the thermal conductivity increases with decreasing porosity in soils without hydrates. Santamarina et al. [73] demonstrated that the general Pythagorean mixing formula is applicable, leading to

$$K_{hbs} = \left[\varepsilon \left(S_h \; K_h^s + S_w \; K_w^s \right) + (1 - \varepsilon) K_m^s \right]^{1/s}.$$

(8)

Note that this expression can be readily extended to include ice and gas phases in cases where these phases may be present. The parallel model corresponds to S=1, and the series model corresponds to S=-1. Adequate predictions for a given soil are obtained with exponents in the $s \approx \pm 0.2$ range. The authors found that, while the thermal conductivity of hydrate is very similar to that of water, marked changes in thermal conductivity occur when high s_h is present in soils.

Santamarina et al. [73] stated that they have emphasized the determination of data trends using physically based mathematical relationships that combine material parameters, instead of merely fitting generic mathematical functions to the data.

It was also noted that the correlation of the measured physical parameters always requires that the hydrate saturation in pore space, which ranges from 0 to 1, be raised to a power greater than 1. This fact significantly reduces the impact of low-hydrate saturations on the measured physical parameters, an effect that is particularly pronounced at the hydrate saturations characteristic of many natural systems (0.2 of pore space). Mechanical properties are strongly influenced by both soil properties and the hydrate loci. Thermal conductivity depends on the complex interplay of a variety of factors, including formation history, and cannot be easily predicted by volume average formulations [73].

CONCLUSIONS

All three main hydrate structures (sI, sII, and sH) are approximately 85% (mol) water and 15% gas when all the cages are filled. This fact suggests that the mechanical properties of the three hydrate structures are similar to those of ice. This conclusion is true to a first approximation, with the exception of thermal conductivity and thermal expansivity. Thermal conductivity of bulk hydrates slightly increases with temperature, contrary to the ice thermal conductivity that decreases with temperature raised to the power of -1.

The physical properties of bulk hydrates are remarkably close to those of pure ice. As a result, the properties of sediments containing hydrate in the pore space are similar to sediments containing normal ice. The morphology of gas hydrates has large effects on sedimentary physical properties, from seismic velocities on a large scale to borehole electrical resistivity on a smaller scale, and, therefore, the gas hydrate morphology impacts the amount of gas hydrate saturation estimated from geophysical data. Hydrate morphology is controlled by the nature of the sediment host as much as by the supply of the necessary chemical components and environmental conditions.

The most readily observable change in sediment physical properties resulting from the formation of gas hydrates is an increase in seismic velocity. Locating likely areas of gas hydrates using remote seismic sensing is relatively straightforward where bottom-simulating reflectors (BSR) are evident. A BSR is a high-amplitude reflector that approximately parallels the seafloor and results from the strong acoustic impedance contrast between the gas-hydrate-bearing sediments above the reflector and the underlying sediments containing free gas. Gas hydrates may be present even where there is no BSR identified from reflection seismic records.

The understanding of the thermal properties of hydrate-bearing sediments is crucial for the future exploitation of methane gas trapped in sediments. In sediment-bearing hydrates, the thermal conductivity reflects the competing effects of the thermal conductivity of the phases involved, their volume fraction, and their spatial distribution. The electrical conductivity is controlled by the availability and mobility of ions. A gradual reduction in conductivity is measured during hydrate formation even though ion exclusion keeps available ions within the unfrozen water. Mechanical properties are strongly influenced by both soil properties and the hydrate loci.

All the physical parameters depend strongly on the hydrate saturation in pore space. This fact significantly reduces the impact of low-hydrate saturations on the measured physical parameters, an effect that is particularly pronounced at the hydrate saturations characteristic of many natural systems (<0.2 of pore space).

ACKNOWLEDGMENTS

Funding for this work was provided by the Department of Energy, National Energy and Technology Laboratory, to Georgia Institute of Technology under Contract no. DE-FC26-06NT42963 and to Prairie View A&M University under Contract no. DE-FG26-06NT42746. Oak Ridge National Laboratory is managed by UT-Battelle, LLC, for the US Department of Energy under contract DE-AC05-00OR22725.

REFERENCES

1. P. Englezos, "Clathrate hydrates," Industrial & Engineering Chemistry Research, vol. 32, no. 7, pp. 1251–1274, 1993.

2. E. D. Sloan Jr., "Gas hydrates: review of physical/chemical properties," Energy & Fuels, vol. 12, no. 2, pp. 191–196, 1998.

3. E. D. Sloan Jr., Clathrate Hydrates of Natural Gases, Marcel Dekker, New York, NY, USA, 3rd edition, 2006.

4. E. G. Hammerschmidt, "Formation of gas hydrates in natural gas transmission Lines," Industrial & Engineering Chemistry, vol. 26, no. 8, pp. 851–855, 1934.

5. Y. F. Makogon, "Hydrate formation in gas bearing strata under perma frost," Gazov Prom-st, vol. 5, pp. 14–15, 1965.

6. K. A. Kvenvolden and T. D. Lorenson, "The global occurrence of natural gas hydrate," in Natural Gas Hydrates, Occurrence, Distribution, and Detection. Geophysical Monograph, C. K. Paull and W. P. Dillon, Eds., vol. 124, pp. 3–18, American Geophysical Union, Washington, DC, USA, 2001.

7. T. S. Collett, "Natural-gas hydrates; resource of the twenty-first century?" Journal of the American Association of Petroleum Geologists, vol. 74, pp. 85–108, 2001.

8. I. Lerche and E. Bagirov, "Guide to gas hydrate stability in various geological settings," Marine and Petroleum Geology, vol. 15, pp. 427–438, 1998.

9. K. A. Kvenvolden, "Gas hydrates—geological perspective and global change," Reviews of Geophysics, vol. 31, no. 2, pp. 173–187, 1993.

10. R. K. McMullan and G. A. Jeffrey, "Polyhedral clathrate hydrates. IX. Structure of ethylene oxide hydrate," The Journal of Chemical Physics, vol. 42, no. 8, pp. 2725–2732, 1965.

11. C. W. Mak and R. K. McMullan, "Polyhedral clathrate hydrates. X. Structure of double hydrate of tetrahydrofuran and hydrogen sulfide," The Journal of Chemical Physics, vol. 42, pp. 2732–2737, 1965.

12. C. A. Koh, "Towards a fundamental understanding of natural gas hydrates," Chemical Society Reviews, vol. 31, no. 3, pp. 157–167, 2002.

13. J. A. Ripmeester, J. S. Tse, C. I. Ratcliffe, and B. M. Powell, "A new clathrate hydrate structure," Nature, vol. 325, no. 6100, pp. 135–136, 1987.

14. J. A. Ripmeester, "The role of heavier hydrocarbons in hydrate formation," in Proceedings of the AIChE Spring Meeting, Houston, Tex, USA, April 1991.

15. A. P. Mehta and E. D. Sloan Jr., "Structure H hydrate phase equilibria of methane + liquid hydrocarbon mixtures," Journal of Chemical & Engineering Data, vol. 38, no. 4, pp. 580–582, 1993.

16. A. P. Mehta and E. D. Sloan Jr., "Structure H hydrate phase equilibria of paraffins, naphthenes, and olefins with methane," Journal of Chemical & Engineering Data, vol. 39, pp. 887–890, 1994.

17. A. P. Mehta and E. D. Sloan Jr., "A thermodynamic model for structure-H hydrates," AIChE Journal, vol. 40, no. 2, pp. 312–320, 1994.

18. A. P. Mehta and E. D. Sloan Jr., "Improved thermodynamic parameters for prediction of structure H hydrate equilibria," AIChE Journal, vol. 42, no. 7, pp. 2036–2046, 1996.

19. R. Sassen and I. R. MacDonald, "Evidence of structure H hydrate, Gulf of Mexico continental slope,"Organic Geochemistry, vol. 22, no. 6, pp. 1029–1032, 1994. ·

20. K. A. Udachin, C. I. Ratcliffe, and J. A. Ripmeester, "A dense and efficient clathrate hydrate structure with unusual cages," Angewandte Chemie, vol. 40, no. 7, pp. 1303–1305, 2001.

21. J. S. Loveday, R. J. Nelmes, M. Guthrie, et al., "Stable methane hydrate above 2 GPa and the source of Titan›s atmospheric methane," Nature, vol. 410, no. 6829, pp. 661–663, 2001.

22. I.-M. Chou, A. Sharma, R. C. Burruss, et al., "Transformations in methane hydrates," Proceedings of the National Academy of Sciences of the United States of America, vol. 97, no. 25, pp. 13484–13487, 2000.·

23. L.Yang, C.A.Tulk, D. D.Klug, et al., "Synthesis and characterization of a new structure of gas hydrate,"Proceedings of the National Academy of Sciences of the United States of America, vol. 106, no. 15, pp. 6060–6064, 2009.

24. D. W. Davidson, "Gas hydrates as clathrates of ices," in Natural Gas Hydrates: Properties, Occurrence and Recovery, J. L. Cox, Ed., pp. 1–16, Butterworths, Boston, Mass, USA, 1983.

25. J. S. Tse, "Dynamicl properties and stability of clathrate hydrates," Annals of the New York Academy of Sciences, vol. 715, pp. 187–206, 1994, 1st International Conference on Natural Gas Hydrates.

26. J. G. Cook and D. G. Leaist, "An exploratory study of the thermal conductivity of methane hydrate,"Geophysical Research Letters, vol. 10, no. 5, pp. 397–399, 1983.

27. R. G. Ross, P. Anderson, and G. Backstrom, "Effects of H and D order on the thermal conductivity of ice phases," The Journal Chemistry Physics, vol. 68, no. 9, pp. 3967–3972, 1978.

28. R. D. Stoll and G. M. Bryan, "Physical properties of sediments containing gas hydrates," Journal of Geophysical Research, vol. 84, no. B4, pp. 1629–1634, 1979.

29. T. Ashworth, L. R. Johnson, and L. P. Lai, "Thermal conductivity of pure ice and tetrahydrofuran clathrate hydrates," High Temperatures-High Pressures, vol. 17, no. 4, pp. 413–419, 1985.

30. J. S. Tse and M. A. White, "Origin of glassy crystalline behavior in the thermal properties of clathrate hydrates: a thermal conductivity study of tetrahydrofuran hydrate," The Journal of Physical Chemistry, vol. 92, no. 17, pp. 5006–5011, 1988.

31. O. Andersson and H. Suga, "Thermal conductivity of normal and deuterated tetrahydrofuran clathrate hydrates," Journal of Physics and Chemistry of Solids, vol. 57, no. 1, pp. 125–132, 1996.

32. A. I. Martin, Hydrate bearing sediments—thermal conductivity, M.S. thesis, School of Civil and Environmental Engineering, Georgia Institute of Technology, Atlanta, Ga, USA, 2004.

33. D. Klaeschen, M. Zillmer, and J. Bialas, "IFM-GEOMAR Report 2002–2004," chapter 3, http://www.ifm-geomar.de/index.php?id=3500.

34. R. A. Duncan, H.C. Larsen, J. F. Allan, et al., "Proceedings of the Ocean Drilling Program," Initial Report164, Ocean Drilling Program, College Station, Tex, USA, 1996.

35. N. Lubick, "Detecting marine gas hydrates," Geotimes, vol. 49, no. 11, pp. 28–30, 2004.

36. T. Yuan, K. S. Nahar, R. Chand, R. D. Hyndman, G. D. Spence, and N. R. Chapman, "Marine gas hydrates: seismic observations of bottom-simulating reflectors off the west coast of Canada and the east coast of India," Geohorizons, vol. 3, no. 1, pp. 1–11, 1998.

37. S. Constable, "Marine electromagnetic methods for gas hydrate characterization," submitted to Seafloor Electromagnetic Methods Consortium, http://marineemlab.ucsd.edu/Projects/GoMHydrate/Proposal_3.pdf.

38. M. Holland, P. Schultheiss, J. Roberts, and M. Druce, "Observed gas hydrate morphologies in marine sediments," in Proceedings of the 6th International Conference on Gas Hydrates (ICGH ‹08), Vancouver, Canada, July 2008.

39. R. Malone, "Gas hydrates topical report," Tech. Rep. DOE/METC/SP-218 (DE85001986), Department of Energy, Morgantown Energy Technology Center, Morgantown, WVa, USA, 1985.

40. A. E. Cook and D. Goldberg, "Gas hydrate filled fracture distribution, eastern Indian continental margin," in Proceedings of the American Geophysical Union. Fall Meeting, San Francisco, Calif, USA, 2007, abstract no. OS11C-04.

41. D. Long, M. A. Lovell, J. G. Rees, and C. A. Rochelle, "Sediment-hosted gas hydrates: new insights on natural and synthetic systems," Geological Society of London, vol. 319, pp. 1–9, 2009.

42. M. B. Clennell, M. Hovland, J. S. Booth, P. Henry, and W. J. Winters, "Formation of natural gas hydrates in marine sediments—part 1: conceptual model of gas hydrate growth conditioned by host sediment properties," Journal of Geophysical Research B, vol. 104, no. B10, pp. 22985–23003, 1999.

43. P. Henry, M. Thomas, and M. B. Clennell, "Formation of natural gas hydrates in marine sediments—part 2: thermodynamic calculations of stability conditions in porous sediments," Journal of Geophysical Research B, vol. 104, no. B10, pp. 23005–23022, 1999.

44. R. L. Kleinberg, C. Flaum, D. D. Griffin, et al., "Deep sea NMR: methane hydrate growth habit in porous media and its relationship to hydraulic permeability, deposit accumulation, and submarine slope stability," Journal of Geophysical Research, vol. 108, p. 2508, 2003.

45. A. P. Camps, Hydrate formation in near surface ocean sediments, unpublished thesis, Department of Geology, University of Leicester, Leicester, UK, 2007.

46. M. Riedel, T. S. Collet, M. J. Malone, and Expedition 311 Scientists, "Expedition 311 summary," inProceedings of the Integrated Ocean Drilling Program (IODP ‹06), vol. 311, Integrated Ocean Drilling Program Management International, Washington, DC, USA, 2006.

47. R. Anderson, M. Llamedo, B. Tohidi, and R. W. Burgass, "Experimental measurement of methane and carbon dioxide clathrate hydrate equilibria in mesoporous silica," The Journal of Physical Chemistry B, vol. 107, no. 15, pp. 3507–3514, 2003.

48. M. Llamedo, R. Anderson, and B. Tohidi, "Thermodynamic prediction of clathrate hydrate dissociation conditions in mesoporous media," American Mineralogist, vol. 89, no. 8-9, pp. 1264–1270, 2004.

49.   U. Tinivella, F. Accaino, M. Giustiniani, and M. F. Loreto, "Gas hydrate and mud volcanoes offshore Antarctic Peninsula: a geophysical study," in Proceedings of the Goldschmidt Conference, Davos, Switzerland, June 2009.

50.   C. Ecker, J. Dvorkin, and A. M. Nur, "Estimating the amount of gas hydrate and free gas from marine seismic data," Geophysics, vol. 65, no. 2, pp. 565–573, 2000.

51.   J. Dvorkin, M. Helgerud, W. Waite, S. Kirby, and A. Nur, "Introduction to physical properties and elasticity models," in Natural Gas Hydrate in Oceanic and Permafrost Environments, M. D. Max, Ed., pp. 245–260, Kluwer Academic Publishers, Dordrecht, The Netherlands, 2000.

52.   G. E. Archie, "The electrical resistivity log as an aid in determining some reservoir characteristics,"Petroleum Transactions of AIME, vol. 146, pp. 54–62, 1942.

53.   E. Spangenberg, "Modeling of the influence of gas hydrate content on the electrical properties of porous sediments," Journal of Applied Geophysics, vol. 56, no. 1, pp. 73–78, 2004.

54.   B. Tohidi, R. Anderson, M. B. Clennell, R. W. Burgass, and A. B. Biderkab, "Visual observation of gas-hydrate formation and dissociation in synthetic porous media by means of glass micromodels," Geology, vol. 29, no. 9, pp. 867–870, 2001.

55.   E. Kingston, Ch. Clayton, and J. Priest, "Gas hydrate growth morphologies and their effect on the stiffness and damping of a hydrate bearing sand," in Proceedings of the 6th International Conference on Gas Hydrates (ICGH ‹08), Vancouver, Canada, July 2008.

56.   G. J. Moridis, T. S. Collett, S. R. Dallimore, T. Satoh, S. Hancock, and B. Weatherill, "Numerical studies of gas production from several CH4 hydrate zones at the Mallik site, Mackenzie Delta, Canada," Journal of Petroleum Science and Engineering, vol. 43, no. 3-4, pp. 219–238, 2004.

57.   A. Koponen, M. Kataja, and J. Timonen, "Permeability and effective porosity of porous media," Physical Review E, vol. 56, no. 3, pp. 3319–3325, 1997.

58.   M. Singh and K. K. Mohanty, "Permeability of spatially correlated porous media," Chemical Engineering Science, vol. 55, no. 22, pp. 5393–5403, 2000.

59. Y. Bernabè, U. Mok, and B. Evans, "Permeability-porosity relationships in rocks subjected to various evolution processes," Pure and Applied Geophysics, vol. 160, no. 5-6, pp. 937–960, 2003.

60. C. H. Arns, M. A. Knackstedt, and N. S. Martys, "Cross-property correlations and permeability estimation in sandstone," Physical Review E, vol. 72, no. 4, Article ID 046304, pp. 1–12, 2005.

61. S. Jin, J. Nagao, S. Takeya, et al., "Structural investigation of methane hydrate sediments by microfocus X-ray computed tomography technique under high-pressure conditions," Japanese Journal of Applied Physics, vol. 45, no. 27, pp. L714–L716, 2006.

62. Y. Jin, J. Nagao, J. Hayashi, et al., "Assessment of the absolute permeability of natural methane hydrate sediments by microfocus X-ray computed tomography," in Proceedings of the 7th ISOPE Ocean Mining Symposium, pp. 93–96, Lisbon, Portugal, July 2007.

63. H. Minagawa, Y. Nishikawa, I. Ikeda, et al., "Relation between permeability and pore-size distribution of methane-hydrate-bearing sediments," in Proceedings of the Offshore Technology Conference, Houston, Tex, USA, May 2008.

64. Y. Yang and A. C. Aplin, "Influence of lithology and compaction on the pore size distribution and modeled permeability of some mudstones from the Norwegian margin," Marine and Petroleum Geology, vol. 15, no. 2, pp. 163–175, 1998.

65. I. N. Tsimpanogiannis and P. C. Lichtner, "Parametric study of methane hydrate dissociation in oceanic sediments driven by thermal stimulation," Journal of Petroleum Science and Engineering, vol. 56, no. 1–3, pp. 165–175, 2007.

66. J. Dvorkin and A. Nur, "Seismic amplitudes from gas hydrates," E&P, November 2007.

67. A. Judge, "Natural gas hydrate in Canada," in Proceedings of the 4th Canadian Permafrost Conference, pp. 320–328, 1982.

68. R. D. Stoll, J. Edwing, and G. M. Bryan, "Anomalous wave velocities in sediments containing gas hydrates," Journal of Geophysical Research, vol. 76, no. 8, pp. 2090–2094, 1971.

69. R.D. Stoll, "Effects of gas hydrates in sediments," in Natural Gases in Marine Sediments, L. R. Kaplan, Ed., pp. 235–247, Plenum Press, New York, NY, USA, 1974.

70. C. Pearson, J. Murphy, and R. Hermes, "Acoustic and resistivity measurements on rock samples containing tetrahydrofuran hydrates: laboratory analogues to natural gas hydrate deposits," Journal of Geophysical Research, vol. 91, no. 14, pp. 14132–14138, 1986.

71. C. Pearson, P. M. Halleck, P. L. McGuire, R. Hermes, and M. Mathews, "Natural gas hydrate deposits: a review of in situ properties," The Journal of Physical Chemistry, vol. 87, no. 21, pp. 4180–4185, 1983.·

72. W. F. Murphy, "Acoustic measures of partial gas saturation in tight sandstones," Journal of Geophysical Research, vol. 89, no. 13, pp. 11549–11559, 1984.

73. J. C. Santamarina and C. Ruppel, "The impact of hydrate saturation on the mechanical, electrical, and thermal properties of hydrate-bearing sand, silts, and clay," in Proceedings of the 6th International Conference on Gas Hydrates (ICGH ‹08), Vancouver, Canada, July 2008.

74. J. C. Santamarina, F. Francisca, T. S. Yun, J. Y. Lee, A. I. Martin, and C. Ruppel, "Mechanical, thermal, and electrical properties of hydrate bearing sediments," in Proceedings of the AAPG Hedberg Research Conference, Vancouver, Canada, September 2004.

75. W. J. Winters, W. F. Waite, D. H. Mason, L. Y. Gilbert, and I. A. Pecher, "Methane gas hydrate effect on sediment acoustic and strength properties," Journal of Petroleum Science and Engineering, vol. 56, no. 1–3, pp. 127–135, 2007.

76. A. Fernandez and J. C. Santamarina, "The effect of cementation on the small strain parameters of sands," Canadian Geotechnical Journal, vol. 38, no. 1, pp. 191–199, 2001.

77. T. S. Yun, J. C. Santamarina, and C. Ruppel, "Mechanical properties of sand, silt, and clay containing tetrahydrofuran hydrate," Journal of Geophysical Research, vol. 112, Article ID B04106, 2007.

78. J. C. Santamarina, K. A. Klein, and M. A. Fam, Soils and Waves: Particulate Materials Behavior, Characterization and Process Monitoring, vol. 488, John Wiley & Sons, New York, NY, USA, 2001.

79. W. B. Durham, L. A. Stern, S. H. Kirby, and S. Circone, "Rheological comparisons and structural imaging of sI and sII end member gas hydrates and hydrate/sediment aggregates," in Proceedings of the 5th International Conference on Gas Hydrates, Tapir Academic, Trondheim, Norway, 2005.

Chapter 4

An Analysis on Stability and Deposition Zones of Natural Gas Hydrate in Dongsha Region, North of South China Sea

Zuan Chen[1], Wuming Bai[1], Wenyue Xu[2], and Zhihe Jin[3]

[1]Key Laboratory of the Study of Earth's Deep Interior, Institute of Geology and Geophysics, Chinese Academy of Sciences, Beijing 100029, China

[2]School of Earth and Atomspheric Sciences, Georgia Institute of Technology, Atlanta, GA 30332, USA

[3]Department of Mechanical Engineering, University of Maine, Orono, ME 04469, USA

ABSTRACT

We propose several physical/chemical causes to support the seismic results which find presence of Bottom Simulating Reflector (BSR) at site 1144 and site 1148 in Dongsha Region, North of South China Sea. At site 1144, according to geothermal gradient, the bottom of stability zone of conduction mode is in agreement with BSR. At site 1148, however, the stability zone of conduction mode is smaller than the natural gas presence zone predicted by the BSR. We propose three causes, that is, mixed convection and conduction thermal flow mode, multiple composition of natural gas and overpressure in deep sediment to explain the BSR presence or gas hydrate presence. Further, our numerical simulation results suggest yet another reason for the presence of BSR at site 1144 and site 1148. Because the temperatures in deep sediment calculated from the mixed convection and conduction thermal flow mode are lower than that from the single conduction mode, the bottom of gas hydrate stability zone (GHSZ) is deeper than the bottom of gas hydrate deposition zone (GHDZ) or BSR. The result indicates that occurrence zone of natural is decided by the condition that natural gas concentrate in the zone is greater than its solubility.

INTRODUCTION

Gas hydrate is an ice-like crystalline mineral in which hydrocarbon and nonhydrocarbon gases are held within rigid cages of water molecules [1]. Geological, geophysical, and geochemical evidence of gas hydrate is reported from 81 localities worldwide onshore in Arctic regions and offshore in passive and active margins and inland seas and lakes [1]. Several recent legs of the Ocean Drilling Program (ODP) have targeted known hydrate locations with the goal of characterizing marine gas hydrate [2]. The data collected from these studies provide a clearer picture of hydrate occurrence on both active and passive continental margins. On the basis of these data, attempts have been made to extrapolate local estimates of hydrate volume to infer a global inventory.

The South China Sea is the largest marginal sea in the western Pacific, and is well known for its abundant oil and gas reserves. A broad and wide continental slope 210,000 km² in area extends to the northern parts of the South China Sea, and is a good site for gas hydrate formation and conservation. Furthermore, a bottom simulating reflector (BSR) has been found on the Northern slope [3].

Site 1144 and Site 1148 which will be discussed in this paper are located in Dongsha region, north of South China Sea and at the lowermost continental slope off southern China (Figure 1). By using seismic, sonic logging, and geothermal data the distribution characteristics of gas hydrate in this zone were studied. BSR and amplitude blanking zone were discovered in seismic profiles. High-velocity interval and velocity reverse were distinguished in sonic logging curve at both sites. As indicated below, GHSZ based on conduction thermal flow model and BSR at site 1144 are basically in agreement, while GHSZ calculated from geothermal gradient using the conduction model is much smaller than the BSR predicated from seismic method at site 1148.

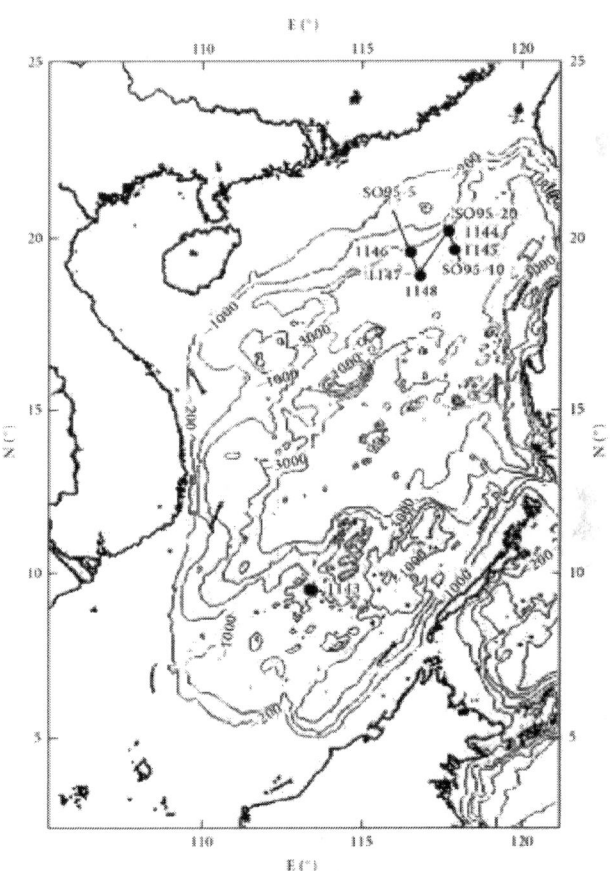

Figure 1: The sit 1148 in Dongsha region north of South China Sea.

In this paper, we aim to explain the results obtained by seismic methods. We first discuss several reasons to explain the difference at site 1148 from view of GHSZ. Second, we perform a numerical analysis to investigate the possible causes, based on a one-dimensional model of gas hydrate formation and evolution in which effects of multiple composition in the gas are considered Third, by comparing the seismic results and simulation analysis, we attempt to find available reason to strengthen our confidence on the presence of gas hydrate at site 1144 and site 1148 in Dongsha region north of South China Sea. In our calculations, based on the mixed convection and conduction thermal flow mode, GHSZ at both sites is found much deeper than GHDZ, which is ensured to be in agreement with BSR.

ANALYSIS ABOUT BOTTOM BOUNDARY OF GHSZ AT SITE 1144

The water depth at site 1144 is 2037 m. The bottom boundary of the gas (gas hydrate stability zone) GHSZ is predicted to be 720 m below sea floor at site 1144 based on the measured thermal gradient of 24∘C /km and a bottom-water temperature of 3.1∘C [3]. Meanwhile, on the basis of data provided by the Ocean Drilling Program (ODP), the presence of gas hydrate in deep-sea sediments can be detected mainly based on the presence of a bottom-simulating reflector (BSR) on seismic profiles, which corresponds to the base of the gas hydrate deposition zone (GHDZ). The BSR depth at site 1144 has been found to be 730 meters below sea floor from seismic profiles, which is in agreement with the result obtained from thermal data [3].

ANALYSIS ABOUT BOTTOM BOUNDARY OF GHSZ AT SITE 1148

The water depth at site 1148 is 3294 m. The thermal gradient is 83∘C /km and the bottom-water temperature is 3.5∘C [3]. Downhole and bottom-water temperature measurements at Site 1148 yielded a thermal gradient of 83∘C /km, which is consistent with the location and water depth [3]. The bottom boundary of the gas (gas hydrate stability zone)

GHSZ is predicted to be 250 m below sea floor at the site based on the temperature (thermal conduction mode), pressure (water depth), and phase equilibrium curve. The BSR depth at site 1148 has been found to be 475 meters below sea floor from seismic profiles, which is not in agreement with the result obtained from thermal data [3]. The possible reasons causing the difference are suggested as follows. (1) Thermal gradient is so big that it is not reliable, which may be excluded according to the reference from Song et al. [3]. (2) The multiple composition in the gas must be considered in the calculation of GHDZ. The temperature at the base of the gas hydrate stability zone may be a few degrees higher than that of methane with pure water. The presence of other gases can make the gas hydrate more stable. For example, ethane and propane contained in the gases have distinguished impact on the stability curve. (3) In a conductive model, thermal gradients remain constant for constant conductivity. Convection, by its nature, tends to increase temperature in the upper part of a system as temperatures in the lower part decrease [4]. It seems more reasonable to assume conduction-convection model rather than a conduction only model for calulating the steady-state temperature distribution. (4) The presence of a deep overpressure along the nearby northern shelf of South China Sea will increase the stability temperature, and a deeper base of GHSZ will be anticipated.

Thermal Gradient

According to the geothermal study, the thermal gradient is 83∘C/km and the water temperature at the sea floor is 3.5∘C at site 1148. The thermal gradient value is much larger than that in the neighbor area. The bottom boundary of the (gas hydrate stability zone) GHSZ is predicted to be 250 m at site 1148 from the geothermal data under hypothesis of pure water and single composition methane, which is much shallower than 475 m obtained from seismic method (Figure 2). It is possible to explain the difference between geothermal study and seismic result by thermal conductive model.

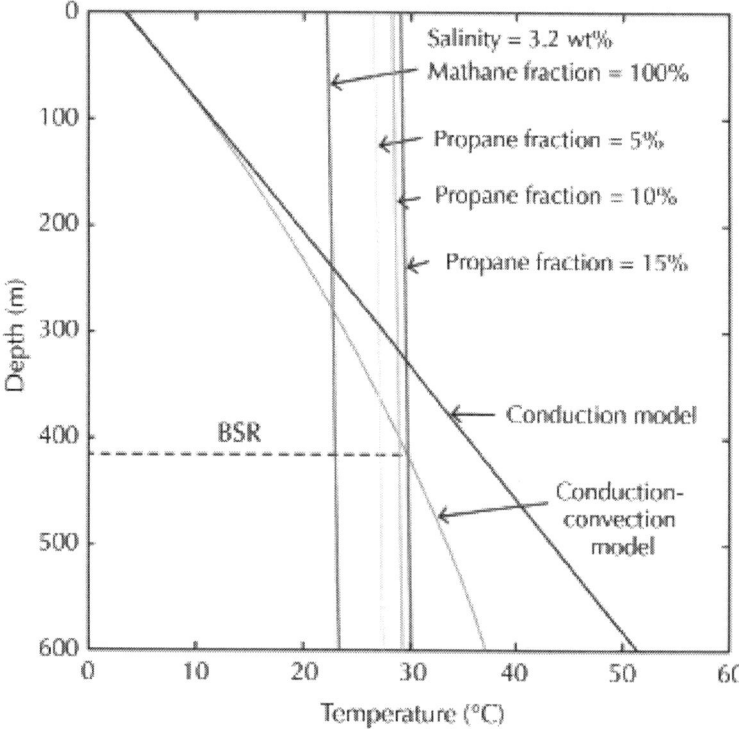

Figure 2: Stability zone predicated from different thermal flow modes and different fraction of methane and propane.

Thermal Flow Mode

The bottom boundary of the GHSZ is determined by comparing the geothermal curve and phase equilibrium curve. The geothermal curve is calculated based on the conduction thermal flow mode in which the temperature is dependent on thermal gradient as shown in Figure 2. Moreover, with a higher total heat flux, the enhanced fluid flow can result in a much lower temperature in deeper sediment while still being consistent with the shallow geothermal data. In the case, thermal flow mode is mixed conduction and convection mode, and the stability boundary will be moved to deeper sediment.

The problem of 1-D steady heat transport due to both fluid flow and heat conduction can be expressed mathematically [5] as follows:

$$\rho_f C_f V_z T - \lambda \frac{dT}{dz} = q_e,$$

$$T(0) = T_0,$$

(1)

where t is the temperature, z is the vertical depth; V_z is the vertical permeation velocity of ground fluid; P_f, C_f, and λ represent fluid density, fluid specific heat and thermal conductivity, respectively; T_0 is the temperature on the top boundary or sea bottom; q_e is constant total heat flux, which can in principle be determined by measuring the temperature, fluid flow velocity, and conductive heat flow at the surface (z = 0).

The solution of this problem is

$$T(z) = T_0 e^{\beta z} + \frac{q_e}{\lambda \beta}\left(1 - e^{\beta z}\right),$$

(2)

Where

$$\beta = \frac{\rho_f C_f V_z}{\lambda}.$$

(3)

Figure 2 shows the calculated geothermal curves under a set of parameters including the vertical permeate velocities of ground fluid and bottom thermal flux. Meanwhile, the stability zone boundaries are obtained under the hypothesis of pure water and single composition methane. Apparently, the bottom boundary of the stability zone is lower than that obtained based on the single conduction mode, which may be the major reason for the discrepancy between the predication of the stability boundary and the anomalies in organic carbon gas contents and geochemical compositions. The gas hydrates containing ethane and propane in addition to methane may also be responsible for the discrepancy.

Multiple Compositions of Natural Gas

As we know, the origins of the gases in hydrate are both thermogenic and biogenic. The gas component from biogenic origin is almost all methane. The gas components from thermogenic origin are manifold. Besides methane, the possible components are ethane, propane, and so on. A lot of research works about thermogenic gas hydrates on the Gulf of Mexico continental slope have been carried out. The chronological framework of the upper 837.11-m composite section (~32.7 Ma to present) at sit 1148 in Dongsha region north of South China Sea has been set up based on biostratiagraphy and magnetostratigraphy [6]. According to the long-term observations of anomalously increased sea-surface temperature scanned by satellite-based thermal infrared and the investigations of the gas geochemistry of bottom water, the authigenic minerals, and the fluid composition, it was concluded that there exist strong degassing and hydrocarbon fluid activities in the submarine. The main composition of submarine gas in the northern continental slopes of the South China Sea is CH_4 and thermogenic [7].

The propane composition has the larger influence on equilibrium temperature comparing with other gas compositions except methane [8]. Hence, two compositions, that is, methane and propane, are considered in our model. Equilibrium temperature of gas hydrate system with both methane and propane increases with pressure and propane fraction, and decreases with salinity of the coexisting liquid solution. The equilibrium temperature is assumed as to be a function of pressure, salinity, and fraction of propane, and is regressed from data obtained using the software CSMHYD [8] as follows:

$$T = \frac{1000}{K} - 273.15,$$

$$K = a_0(h,s) + a_1(h,s)\log(p) + a_2(h,s)\log(p)^2$$
$$+ a_3(h,s)\log(p)^3 + a_4(h,s)\log(p)^4,$$

$$a_1(h,s) = b_{i0}(s) + b_{i1}(s)h + b_{i2}(s)h^2$$
$$+ b_{i3}(s)h^3 + b_{i4}(s)h^4, \quad (i = 0,1,2,3,4),$$

(4)

Where T is temperature, p is pressure, h is fraction of propane in natural gas, and s is salinity.

Figure 2 indicates that phase equilibrium curves change with the fraction of propane. Apparently, the stability boundary will be moved to deeper sediment, and the change is obvious. When propane fraction reaches 10%, the depth of bottom stability boundary is 475 m, which agrees with the BSR result. Moreover, the hypothesis that propane and other gas composition exist in natural gas in addition to methane in site 1148 is strongly supported by the measurements of organic carbon gases, which may indicate hydrate forming gases carried by the fluid flow from a deep source [9].

Overpressure under Deep Sediment Region

Deposit environment in the north aktian zone of South China Sea is mainly shore, infraneritic, and bathyal region. The emptied terrigenous material is rich, and the laid down velocity is high. The condition is very prominent after Miocene epoch. Therefore, a deep overpressure by rapid sedimentation along the nearby northern shelf of the South China Sea can be another cause for enhancing phase equilibrium temperature and a high upward fluid flow. Of course, we do not have direct insite measured data to support the hypothesis. If the convection thermal flow mode at site 1148 is possible, it will be indirect evidence to indicate overpressure existence in the region. Moreover, the stability temperature only increases 3.5°C when the pressure increases 10 MPa. That means overpressure in deep sediment is not a major factor affecting the boundary of hydrate deposition.

NUMERICAL SIMULATION OF GHSZ AND GHDZ AT SITE 1144 AND SITE 1148

To carry out numerical simulation for the dynamic process of hydrate in marine sediment, we need (1) to create phase-inversion model of natural gas hydrate, which concerns the pressure, temperature, and salinity of the fluid, as well as phase equilibrium condition, solubility, fluid density, and enthalpy. The simulation results would be unreliable if the dynamic process of hydrate phase-inversion was not considered precisely, and (2) to create a general mathematical model which can describe the transport process of fluid, heat, natural gas, and salt, as well as some geological processes, such as sedimentation and natural gas formation resource.

Stability zone and occurrence zone of methane hydrate formation can be predicted under certain boundary conditions using the calculation model of methane hydrate in seafloor sediment presented by Xu et al. [10–12]. Stability zone is decided by intersection of phase transition curve and actual ground temperature curve. Gas hydrate occurrence zone must lie in stability zone, and is decided by the condition that gas concentration is greater than its solubility. So prediction of stability zone and occurrence zone of gas hydrate formation will be effected by the distribution of actual temperature and pressure as well as mass flow and gas flow at the boundary.

The phase transition process, that is, the phase equilibrium relationship of hydrate formation process, was considered by Xu et al. [11, 12], and Chen et al. [13], and the formation or dissolution was described dynamically through enthalpy change. The calculation method of phase transition in this paper is the same as that used in Xu et al. [11, 12] and Chen et al. [13]. In fact, natural gas with certain proportion of compositions usually changes their proportions in hydrate after phase transition, that is, composition proportion of natural gas will also change. We do not take into account the complex process and assume simply that composition proportion of natural gas is fixed.

Model Formulation

The mathematical model formulation used in the calculation is summarized as follows [11–13].

Assuming natural gas diffusion occurs only in the liquid phase, the transport of natural gas can be described by

$$\frac{\partial(\phi \rho C)}{\partial t} + \frac{\partial}{\partial z}\left(q_l C_l + q_g C_g + q_h C_h - \phi S_l \rho_l D_c \frac{\partial}{\partial z} C_l\right) = Q, \tag{5}$$

where t is the time, D_c is the diffusivity of natural gas in the liquid solution, Q represents the rate of in situ natural gas production, subscripts l, g, and h represent liquid, gas, and hydrate, respectively, φ is porosity; ρ, C, and q are the density, natural gas concentration, and Darcian flow rate, respectively; S is the saturation, and z is the spatial coordinate pointing upward.

$$q_l = -\frac{kk_l\rho_l}{\mu_l}\left(\frac{\partial}{\partial z}p + \rho_l g\right),$$

$$q_g = -\frac{kk_g\rho_g}{\mu_g}\left(\frac{\partial}{\partial z}p + \rho_g g\right),$$

$$(6)$$

Where g is the gravitational acceleration, k is the permeability of the porous medium, P is the pressure, μ is the fluid viscosity, and kl and kg are relative permeability of liquid and gas.

Conservation of energy is cast in terms of enthalpy H and temperature T and written as

$$\frac{\partial}{\partial t}[\phi\rho H + (1-\phi)\rho_s H_s] + \frac{\partial}{\partial z}\left[q_l H_l + q_g H_g + q_h H_h - \lambda\frac{\partial}{\partial z}T\right]$$

$$= 0,$$

$$(7)$$

Where subscript s refers to the sediment matrix, C_s denotes specific heat capacity, and λ is effective thermal conductivity

Numerical Calculation of GHSZ and GHDZ around Site 1144

Seafloor sediment with 800 m thickness is accounted, and 401 junction points are marked. Table 1 lists the physical parameters of the medium. The boundary conditions are as follows. Pressure in seafloor is 20.37 MPa, temperature in seafloor is 3.1°C, weight percentage of natural gas content in seafloor is 10^{-6}%, weight percentage of salt in seafloor is 3.2%, heat flow in the bottom of computed zone is 0.03 W/m², mass flow in bottom is 2×10^{-7} kg/m²/s, natural gas flow in bottom is 3.035×10^{-10} kg/m²/s, and salt flow in bottom 3.2×10^{-9} kg/m²/s. The natural gas is pure methane. The above parameter values have eventually been determined through trials. The calculated GHSZ and GHDZ are shown in Figure 3, in which z is depth from seafloor; bottom abscissa means temperature T, top abscissa is natural gas content C. The results indicate that when the mass flow in bottom is large enough, the GHDZ will exist in sediment and its bottom depth 720 m will be consistent with BSR from seismic method. In this case, a mixed conduction and convection thermal flow mode is used, and the stability boundary will be moved to deeper sediment than the bottom boundary of GHDZ.

Table 1

Parameter	Value
g	9.81 ms-2
φ	0.5
k	1×10-16 m2
ρh	920.0 kgm-3
λ	1.0 Wm-1C-1
Cs	1000.0 Jkg-1C-1
μl	0.888×10-3 kgm-1s-1
Dc	1.0×10-9 m2s-1

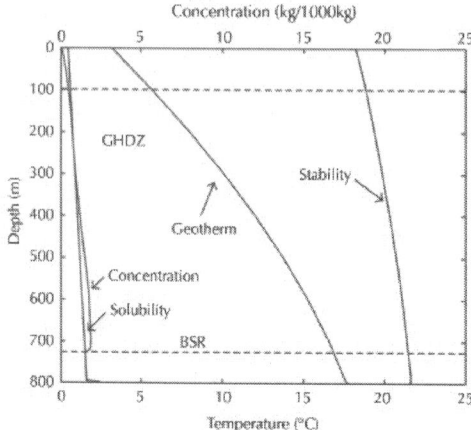

Figure 3: Numerical simulation for GHSZ and GHDZ at site 1144.

Numerical Calculation of GHSZ and GHDZ around Site 1148

Seafloor sediment with 600 m thickness is accounted, and 301 junction points are marked. Boundary conditions are pressure in seafloor of 32.94 MPa, temperature in seafloor is 3.5°C, weight percentage of natural gas content in seafloor is 10^{-6}%, weight percentage of salt in seafloor is 3.2%, heat flow in bottom of computed zone is 0.06 W/m², mass flow in bottom is 1.5×10^{-7} kg/m²/s, natural gas flow in bottom is 3.035×10^{-10} kg/m²/s, and salt flow in bottom is 3.2×10^{-9} kg/m²/s. The natural gas is composed of methane (88%) and propane (12%). Through many times trial, the parameter values mentioned above have eventually been determined. The calculated GHSZ and GHDZ are shown in Figure 4. The results indicate that when the mass flow in bottom is large enough, the GHDZ will exist in sediment and its bottom depth 475 m will be consistent with BSR from seismic method. Again, a mixed conduction and convection thermal flow mode is used, and the stability boundary will be moved to deeper sediment than the bottom boundary of GHDZ just as site 1144.

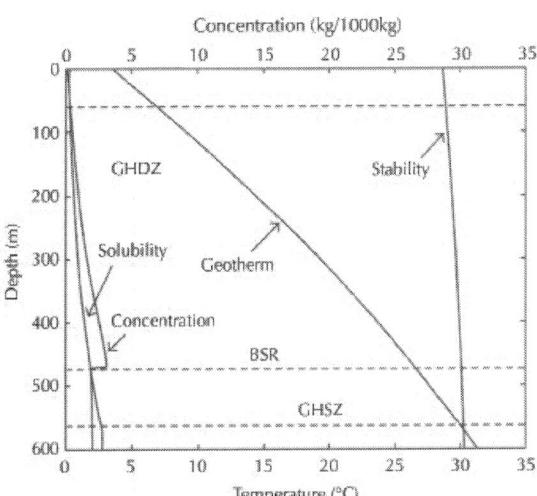

Figure 4: Numerical simulation for GHSZ and GHDZ at site 1148.

CONCLUSIONS

In this paper, we propose several possible reasons to support the seismic method results which find presence of BSR at site 1144 and site 1148. At site 1144, according to geothermal gradient, the bottom of stability zone of conduction mode is in agreement with BSR or 720 m depth bellow seafloor. At site 1148, due to the high geothermal gradient, the stability zone of conduction mode is less than the natural gas presence zone predicted by BSR which is 475 m below seafloor. BSR presence or gas hydrate presence may be explained by the following three mechanisms, that is, mixed conduction and convection thermal flow mode, multiple composition in natural gas, and overpressure in deep sediment. Logging result in the region [3] (Song, 2001) also indicate that bottom boundary of GHSZ at site 1148 is at 475 m. We numerically simulate the dynamic process of gas hydrate evolution at site 1144 and site 1148, and GHDZ will reach 720 m and 475 m, respectively. Moreover, numerical results also indicate that in the two cases, thermal flow is mixed conduction and convection mode, and GHSZ is deeper than GHDZ. These results also indicate that GHSZ is a basic condition for presence of GHDZ, which is finally predicted by the mathematical model considering the dynamic process of hydrate in marine sediment.

It should be pointed out that our conclusions are only theoretical inference in order to explain the seismic results. Real situations at site 1144 and 1148 need more direct investigation.

ACKNOWLEDGMENTS

The authors would like to thank two anonymous reviewers and the associated editor for their critical but helpful comments. The research work is supported by CNSF (40274026, 40874046).

REFERENCES

1.   E. D. Sloan Jr., "Fundamental principles and applications of natural gas hydrates," Nature, vol. 426, no. 6964, pp. 353–359, 2003.

2. C. K. Paull, R. Matsumoto, and P. Wallace, "Leg 164 science party," in Proceedings of the Ocean Drilling Program, vol. 164 of Initial Reports, College Station, Tex, USA, 1996.

3. H.-B. Song, J.-H. Geng, H.-K. Wang, et al., "A preliminary study of gas hydrates in Dongsha Region North of South China Sea," Chinese Journal of Geophysics, vol. 44, no. 5, pp. 687–795, 2001 (Chinese).

4. R. W. Rex and D. J. Howell, "Assessment of U.S. geothermal resources," in Geothermal Energy, P. Kruger and C. Otte, Eds., 1973.

5. J. D. Bredehoeft and I. S. Papadopulos, "Rates of vertical groundwater movement estimated from the earth's thermal profile," Water Resources Research, vol. 1, no. 2, pp. 325–328, 1965.

6. P. Wang, W. L. Prell, P. Blum, et al., "Leg 184," in Proceedings of the Ocean Drilling Program, vol. 184 ofInitial Reports, College Station, Tex, USA, 2000.

7. B. Wu, G. Zhang, Y. Zhu, et al., "Progress of gas hydrate investigation in China offshore," Earth Sciences Frontiers, vol. 10, no. 1, pp. 177–189, 2003 (Chinese).

8. E. D. Sloan Jr., Clathrate Hydrates of Natural Gases, Marcel Dekker, New York, NY, USA, 1998.

9. Y. Zhu, Y. Huang, M. Ryo, and B. Wu, "Geochemical and stable isotopic composition of pore fluids and authigenic siderite concretions from site 1146, ODP leg 184: implications for gas hydrate," in Proceedings of the Ocean Drilling Program, Scientific Results, W. L. Prell, P. Wang, P. Blum, D. K. Rea, and S. C. Clemens, Eds., vol. 184 of Initial Reports, pp. 1–15, 2000.

10. W. Xu and C. Ruppel, "Predicting the occurrence, distribution, and evolution of methane gas hydrate in porous marine sediments," Journal of Geophysical Research B, vol. 104, no. 3, pp. 5081–5095, 1999.

11. W. Xu, R. P. Lowell, and E. T. Peltzer, "Effect of seafloor temperature and pressure variations on methane flux from a gas hydrate layer: comparison between current and late Paleocene climate conditions,"Journal of Geophysical Research B, vol. 106, no. 11, pp. 26413–26423, 2001.

12. W. Xu, "Modeling dynamic marine gas hydrate systems," American Mineralogist, vol. 89, no. 8-9, pp. 1271–1279, 2004.

13. Z. Chen, W. Bai, and W. Xu, "Prediction of stability zones and occurrence zones of multiple composition natural gas hydrate in marine sediment," Chinese Journal of Geophysics, vol. 48, no. 4, pp. 936–945, 2005 (Chinese).

Chapter 5

Assessment of Marine Gas Hydrates and Associated Free Gas Distribution Offshore Uruguay

Juan Tomasini[1], Héctor de Santa Ana[1], Bruno Conti[1], Santiago Ferro[1], Pablo Gristo[1], Josefina Marmisolle[1], Ethel Morales[1], Pablo Rodriguez[1], Matías Soto[1], and Gerardo Veroslavsky[2]

[1]ANCAP, Exploración y Producción, Paysandú s/n y Avenida Del Libertador, 6to piso, 11100 Montevideo, Uruguay

[2]Departamento de Evolución de Cuencas, Facultad de Ciencias, Instituto de Ciencias Geológicas, Iguá 4225, 11400 Montevideo, Uruguay

ABSTRACT

Natural gas hydrates are crystalline solids formed by natural gas (mainly methane) and water that are stable under thermobaric conditions of high pressure and low temperature. Methane hydrate is found in polar areas of permafrost and in offshore basins of continental margins.

These accumulations may represent an enormous source of methane. Based on global estimations of methane concentration in natural gas hydrates, the methane content may be several times greater than those of technically recoverable, conventional natural gas resources. In the continental margin of Uruguay, seismic evidence for the occurrence of gas hydrate is based on the presence of (bottom simulating reflectors) BSRs in 2D seismic reflection sections. Here we present results regarding gas hydrates and associated free gas distribution assessment offshore Uruguay, based on BSR mapping and applying a probabilistic approach. A mean value of 25,890 km² for the area of occurrence shows a great potential for this nonconventional resource, encouraging further research.

INTRODUCTION

Natural gas hydrates are crystalline solids formed by natural gas (mainly methane) and water that are stable under thermobaric conditions of high pressure and low temperature [1].

Methane hydrate occurs in sediments within and below thick permafrost in Arctic regions and in the subsurface of most continental margins where water depths are greater than 500 meters [2].

Gas hydrate accumulations may represent an enormous source of methane. Based on global estimations of methane concentrations, the methane content is about 2 to 10 times greater than those of technically recoverable conventional natural gas resources [2]. The existence of such a large methane hydrate resource has provided a strong global research incentive and international interest, which has grown in the last years.

The first acoustic indication of gas hydrate occurrence is given by presence of (bottom simulating reflection) BSR in seismic sections due to a significant change in acoustic impedance between sediment containing hydrates and sediments containing free gas [3, 4]. The seismic appearance is parallel to seafloor reflection with a polarity reversal with respect to the seafloor.

The BSR is usually a good indication of gas trapped below the base of the gas hydrate stability zone (GHSZ) implying that gas hydrates are present [1]. On the other hand, gas hydrate can exist without creating a well-defined BSR, especially when gas fluxes are directed through faults or comparable permeable fluid pathways [1].

In offshore basins around the world the base of the GHSZ can have different seismic expressions such as continuous, segmented, and high-relief BSRs depending on the stratigraphic, fluid, and geothermal setting [5].

Another seismic response associated with the presence of gas hydrates in marine sediment is the blanking (reduction of the amplitude of seismic reflections). It can be used to identify sediments, in which hydrates have been formed. However, blanking is not a good indicator of the base of GHSZ because there are several possibilities leading to signal attenuation, like the original or diagenetic character of strata as well as artifacts produced during seismic processing [1].

The study area for this work is the continental margin of Uruguay. This margin was formed during continental rifting and seafloor spreading, which included strong volcanic activity [6]. Three offshore basins were created during this process: Punta del Este, Pelotas, and Oriental del Plata basins (Figure 1) [7], which have a total extent, within the 200 nautical miles limit, of near 85,000 km² and a maximum volcano-sedimentary fill of 8,000 m based on seismic data [7]. This volcano-sedimentary fill comprises Juro-Cretaceous to recent sequences and is worth to mention the presence of a Paleozoic prerift correlative with Parana Basin records.

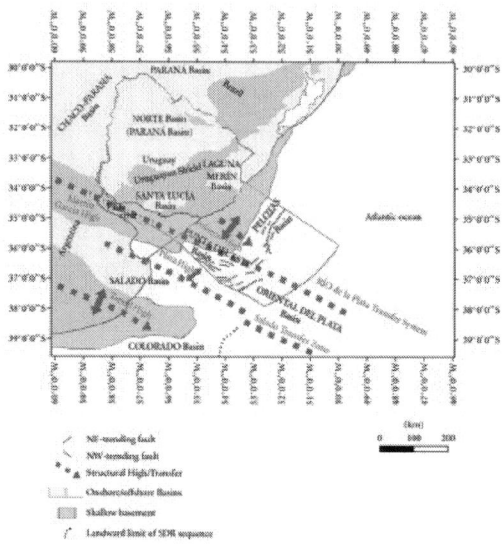

Figure 1: Sedimentary basins of Uruguay. After Soto et al. [7].

These basins are genetically related to the Western Gondwana breakup (~130 Ma ago) and the subsequent development of the Atlantic Ocean and thus, are part of an important series of depocenters which include offshore hydrocarbon productive basins such as Santos and Campos basins (Brazil) and also the conjugate Orange Basin (South Africa and Namibia) [7].

The Punta del Este Basin is a mainly NW-SE trending aborted rift, perpendicular to the general trend of the continental margin [8]. In contrast, the NE-SW trending Pelotas Basin belongs to the flexural border of a precursor rift structure and continues in the Brazilian margin up to the Florianópolis Platform [7].

The Punta del Este and Pelotas basins are separated in shallow waters by the Polonio High. The distal part of both basins, where the Polonio High is not present and comprises a thick Cenozoic package, is called by some authors the Oriental del Plata Basin (Figure 1) [9, 10], which according to a recent redefinition has been restricted to the southern region of the continental margin, comprising only transitional and oceanic crust [7].

Different water masses and currents coexisting in the area play a fundamental role in the occurrence of gas hydrates considering temperature, salinity, and pressure conditions as well as sediment erosion and deposition.

Today, the continental margin of Uruguay is characterized by strong contour currents and the important input of huge amounts of sediments from the Río de la Plata river [11].

The area comprises a very complex and dynamic oceanographic regime. At surface level, dense and cold antarctic water masses from the Malvinas/Falkland Current flowing northward converge with the warm and saline Brazil Current flowing towards the South, resulting in the Brazil-Malvinas Confluence [12].

However, the confluence is not confined to surface currents, and also the interaction of intermediate water masses results in a complicated flow pattern. While antarctic intermediate water (AAIW) and circumpolar deep water (CDW) are flowing northward, the southward flowing north atlantic deep water (NADW) separates the CDW into Upper-CDW and Lower-CDW. The deep basins are under the influence of the antarctic bottom water (AABW) [13].

Interaction between these currents strongly affects sedimentary processes as well as margin morphology. The existence of strong contour currents leads to the generation of a large Contouritic Depositional Complex, which at least extends from southern Argentine margin to the margin of Uruguay, including various kinds of erosive and depositional sedimentary features [14].

Although strong deep currents are not favorable for organic matter deposition and preservation, contouritic deposits have been frequently associated both with conventional hydrocarbon reservoirs and with gas hydrate accumulations [15]. Indeed, big hydrate accumulations inferred from BSRs have been found in contouritic deposits along Atlantic margin [15].

In addition, these along-slope processes interact with downslope sedimentary gravitational processes, which have a large impact in the study area. In this way, mainly in the southern region (Punta del Este Basin), a series of submarine channels are developed.

Conventional hydrocarbon reservoirs associated with turbiditic sequences are well known while methane hydrate accumulations in this type of sequences have been reported as drilling targets offshore Japan [16].

First work regarding gas hydrates offshore Uruguay was performed by de Santa Ana et al. [17], where the presence of a BSR was recognized for the first time. Gas hydrate distribution and thickness were afterwards estimated based on the available seismic grid allowing first approximations on resource potential. Initial determination of mineralized area was 5,000 km^2 resulting in 86 trillion cubic feet (TCF) of natural gas under normal conditions [17], based on seismic information available at that time in nondigital format.

In 2005, the presence of gas hydrates was reported by Neben et al. from the German institute BGR after a 2D seismic survey in the area [18]. In this work, the BSR area was mapped from seismic sections acquired at that survey, resulting in a minimum of approximately 7,000 km^2.

Even if the BSR represents the most reliable indication of the existence of gas hydrates within the study area, high methane concentrations and (anaerobic oxidation of methane) AOM within the upper few meters of the sediments suggest the existence of methane hydrate in the study area [19].

Hydrocarbon generation and migration offshore Uruguayn have been confirmed through fluid inclusion analysis [20], which were recognized in synrift and postrift sequences from two wells drilled in the area [21].

In 2008, oil seeps were identified by satellite images [10, 22], and poststack processing for gas chimney identification [23] was performed on 2D seismic sections, which showed vertical disturbances of the seismic signal. These signal anomalies, reaching into higher depths, were interpreted as hydrocarbon migration pathways and suggest a thermogenic origin of the gases that reach gas hydrate reservoirs [24].

All these evidences of hydrocarbon generation strongly indicate that thermogenic provenance of gas to form gas hydrates in upper sequences, as the only or combined with biogenic gas origin, is very plausible.

In this work, the integration of a dense grid of different 2D seismic surveys acquired offshore Uruguay was done for first time in order to identify the base of the GHSZ and assess gas hydrate and associated free gas distribution within the studied area. A probabilistic approach was considered in order to reflect the uncertainty of the interpretation, taking into account both high and low side of the mapped area and reporting a final mean value.

METHODOLOGY

For this work we used 2D reflection seismic data, acquired during different surveys for hydrocarbon exploration offshore Uruguay between 1970 and 2008.

More than 24,000 km of 2D reflection seismic sections were interpreted (Table 1). Seismic-stratigraphic interpretation was performed using commercial seismic interpretation software, considering reflectors attributable to the base of GHSZ, based on its theoretical response (Figure 2) and known characteristics of BSRs [1, 3, 4].

Table 1: Offshore Uruguay seismic survey's details

Survey	Acquisition company	Year	Number of lines	Length (km)
UR70	CGG	1970	12	2571
UR71	CGG	1971	32	2696
UR74	GSI	1974	35	2578
UR75	GSI	1975	28	1897
UR77	GSI	1977	16	1050
UR82	WESTERN	1982	23	1402
UR02	CGG	2002	6	1850
UR07	WAVEFIELD-INSEIS	2007	32	7125
UR08	WAVEFIELD-INSEIS	2008	22	2909

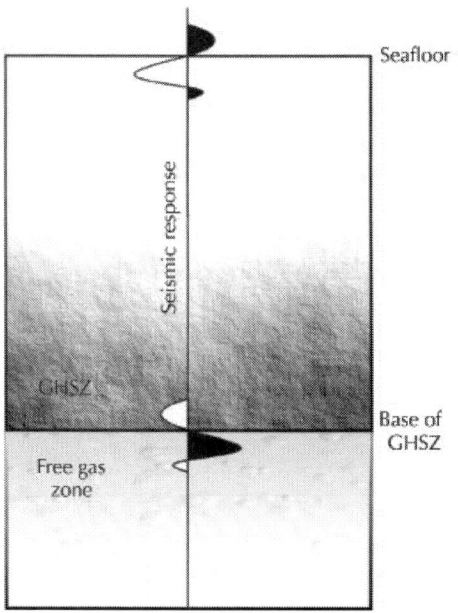

Figure 2: Theoretical response of seafloor and base of GHSZ.

Once seismic interpretation was performed on available sections, gas hydrate and free gas spatial distribution was mapped and calculated.

RESULTS

Interpretation of the base of GHSZ from seismic data in the area shows a widespread distribution of gas hydrate bearing sediments.

Continuous and segmented BSRs were observed while "high relief" BSRs were not identified. In Figure 3 we present an example of seismic wiggle showing the characteristic polarity reverse of BSR regarding the seafloor. For some sections, acoustic blanking is observed above a BSR cross-cutting sedimentation strata (Figure 4). In Figure 5 we present a seismic section from UR08 survey shot at Punta Del Este Basin; a clear example of BSR is observed while seismic anomalies of enhanced amplitudes are present below it.

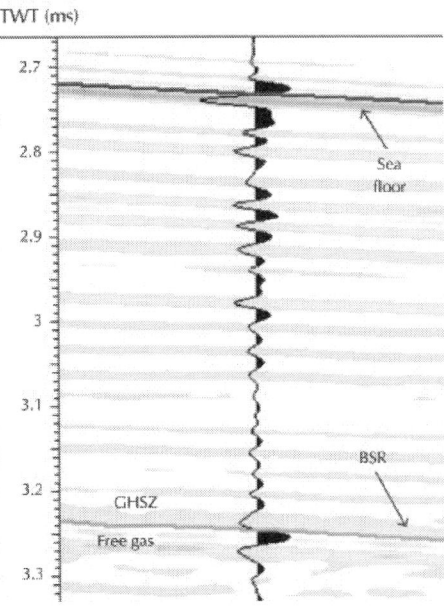

Figure 3: Seismic response of seafloor and base of GHSZ in seismic section from the area.

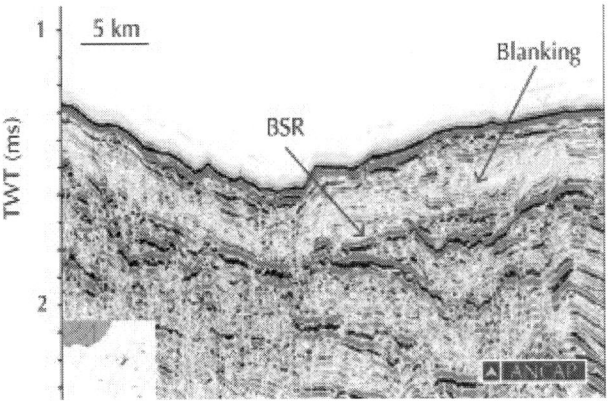

Figure 4: Section UR82_004 from Pelotas Basin. BSR present at 1982 survey showing blanking at the hydrate zone.

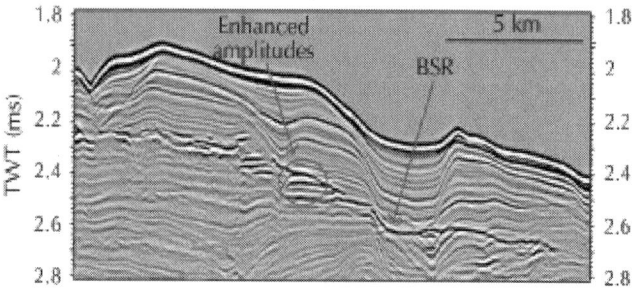

Figure 5: Seismic section from Punta del Este Basin showing a BSR at 0.330 sec TWT from the seafloor and enhanced amplitudes bellow the BGHSZ. Modified from [24]. Courtesy of CGG Veritas.

BSR was interpreted in UR70, UR71, UR74, UR77, UR82, UR02, UR07, and UR08 surveys. It is present below water depths from 500 to 3,200 m and has high continuity in Pelotas Basin but is more discontinuous at Punta del Este Basin (Figure 6).

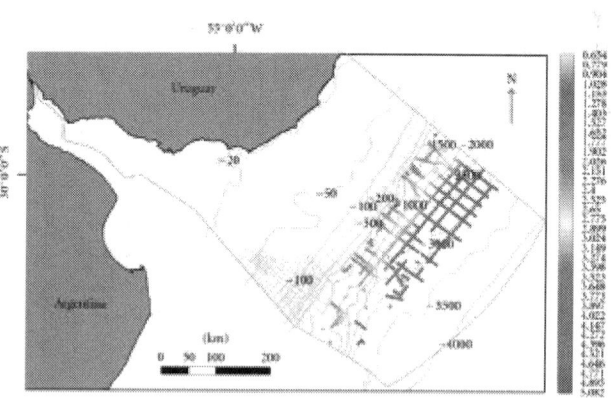

Figure 6: Results of BSR interpretation offshore Uruguay, showing a widespread distribution of gas hydrate bearing marine sediments. Color scale represents reflector depth in ms (TWT).

BSR interpretations showed in Figure 6 corresponds to 25,000 km^2 and for probabilistic calculation purposes is considered as the P50 of the occurrence area.

For the high side, we have mapped the total maximum area of sediments that may contain gas hydrates offshore Uruguay considering the envelope of BSR interpretations. The total maximum area corresponds to 32,500 km² (show in Figure 7) and was considered as a P5 percentile ("larger than" convention).

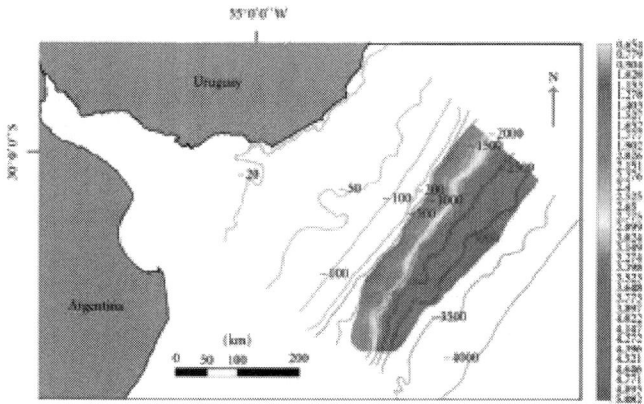

Figure 7: Maximum area of gas hydrate and free gas distribution, considering the envelope of BSR interpretations.

The distribution chosen for the area assessment was of log-normal type as being the most common followed by nature and in particular for resource area [25]. Considering μ=Ln(P50) and truncation for values greater than P1 in order to disregard the extremely high values of the distribution, the resulting mean value for the area is 25.89 × 10³ km² as shown in Figure 8.

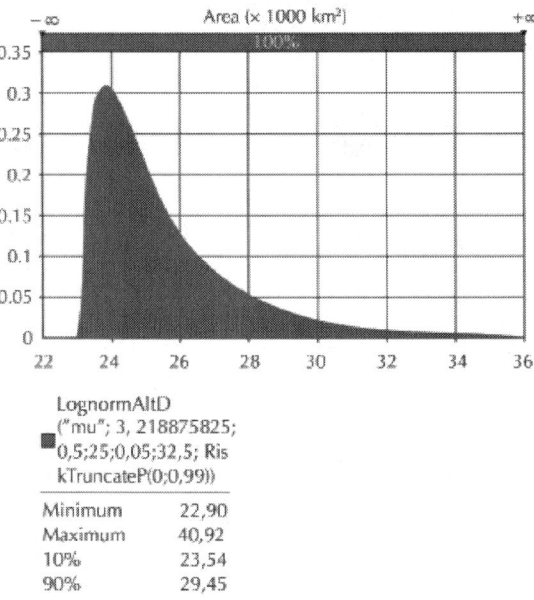

LognormAltD
("mu"; 3, 218875825;
0,5;25;0,05;32,5; Ris
kTruncateP(0;0,99))

Minimum	22,90
Maximum	40,92
10%	23,54
90%	29,45

Figure 8: Probability density function of the lognormal distribution of gas hydrates occurrence area offshore Uruguay. P90 = 23.54 × 10^3 km²; P10 = 29.45 × 10^3 km².

DISCUSSION

In Figure 4, acoustic blanking is observed above BSR. As showed by Max et al. [1], this seismic character may have a relation with gas hydrate presence. In this case, the fact of finding blanking associated with BSR supports the interpretation of hydrate occurrence. In Figure 5, seismic anomalies of enhanced amplitudes suggests the presence of free gas below the base of GHSZ opening new opportunities for further research on subhydrate prospects exploration in the area.

At basin scale, as water depth increases, the base of the GHSZ progressively extends farther below the seafloor. This increment in thickness of GHSZ responds to pressure increasing and temperature decreasing at higher water depths as showed by Max et al. [1]. As reported for several offshore basins around the world [2], gas hydrate occurrence offshore Uruguay starts developing at approximately

500 m of water depth (Figure 6). This agrees with the P-T seafloor conditions for phase equilibrium [26] of approximately 4903 KPa, 5°C, and seawater salinity of 35 PSU for a CH_4 + NaCl + H_2O system [27]. Further studies are needed in order to assess possible gas hydrate formation and destabilization cycles with eustasy and its effects on seafloor stability and geohazards potential.

Towards NE direction, BSR interpretation stops at or close to the end of strike lines, as the available database does not cross the Uruguay-Brazil maritime boundary. Nevertheless, interpreted gas hydrate area offshore Uruguay is probably part of a much greater gas hydrate province, shared with Brazil, extending further north at the Pelotas Basin. This basin has the largest gas hydrate occurrence of the Brazilian coast [28, 29].

At SE direction, BSR was interpreted up to a maximum water depth of 3,200 m. Below this ultradeep water domains, parallelism of sedimentary reflectors makes the identification of BSR difficult. Nevertheless, gas hydrate may be present beyond this range of water depth as found at other locations around the world and reported by Booth et al. [30].

Distribution of gas hydrate deposits offshore Uruguay presents high continuity in the north area (PelotasBasin) but is more discontinuous in the Punta del Este Basin (while it cannot be assessed in the Oriental del Plata Basin due to limited seismic data; Figure 6). This may be due to various factors, including the presence of submarine channels. The existence of this kind of morphologies associated with sediment transport processes down slope has a special importance in Punta del Este Basin and may be the cause of BSR interpretation absence in certain zones, mainly due to complex sedimentary disposition.

In addition, the base of GHSZ has different expressions depending on lithological characteristics of sediments, leading to cases where the BSR may not be evident. In these cases, different relative sand-clay contents may play an important role.

On the other hand, the presence of BSRs indicates the coexistence between Free Gas and Gas Hydrate phases therefore situations may exist where sediments contain hydrates but not enough free gas at the phase boundary as to generate a BSR-like seismic response.

All these different aspects regarding BSR imaging and interpretation support the importance of the probabilistic approach applied to report area of occurrence.

CONCLUSIONS

Gas hydrate and associated free gas occurrence presents a widespread distribution offshore Uruguay, resulting in a mean value of $25,890\,km^2$, being much higher than reported by previous studies. Results show a great potential for this nonconventional resource, encouraging further research.

Further studies are needed about possible gas hydrate formation and destabilization cycles that may take place at the landward limit of the occurrence zone in order to evaluate its effect on seafloor stability and geohazard potential.

Seismic evidence of discrete free gas accumulations below the gas hydrate stability zone through amplitude anomalies was found. Those accumulations could be considered as subhydrate prospects and may play an important role considering future gas field developments offshore Uruguay.

From the exploratory point of view, determination of locations with good reservoir characteristics is critical for a comprehensive resource assessment.

Identification and quantification of high porosity and permeability sand deposits within the mapped area and GHSZ is needed in order to define exploratory targets. This will be the key element for the eventual development of gas hydrate prospects offshore Uruguay, once international research manages to prove that methane from marine gas hydrates can be produced as a technically safe, environmentally compatible, and economically competitive energy resource.

ACKNOWLEDGMENTS

This work was performed in the framework of the Project FSE_2009_53 from the ANII. CGGVeritas and ANCAP are acknowledged for the permission to publish figures of seismic sections.

REFERENCES

1. M. D. Max, A. H. Johnson, and W. P. Dillon, Economic Geology of Natural Gas Hydrate, Springer, 2006.

2. Committee on Assessment of the Department of Energy Methane Hydrate Research and Development Program: Evaluating Methane Hydrate as a Future Energy Resource, Realizing the Energy Potential of Methane Hydrate for the United States, 2010.

3. E. D. Sloan, Clathrate Hydrates of Natural Gas, Marcel Dekker, New York, NY, USA, 1998.

4. C. K. Pecher and W. S. Holbrook, "Seismic methods for detecting and quantifying gas hydrates," inNatural Gas Hydrate in Oceanic and Permafrost Environments, M. D. Max, Ed., pp. 257–294, Kluwer Dordrecht, 2000.

5. B. Shedd, P. Godfriaux, M. Frye, R. Boswell, and D. Hutchinson, Occurrence and Variety in Seismic Expression of the Base of Gas Hydrate Stability in Gulf of Mexico, USA, Fire in The Ice, Methane Hydrate Newsletter, 2009.

6. D. Franke, S. Neben, S. Ladage, B. Schreckenberger, and K. Hinz, "Margin segmentation and volcano-tectonic architecture along the volcanic margin off Argentina/Uruguay, South Atlantic," Marine Geology, vol. 244, no. 1–4, pp. 46–67, 2007.

7. M. Soto, E. Morales, G. Veroslavsky, H. de Santa Ana, N. Ucha, and P. Rodríguez, "The continental margin of Uruguay: Crustal architecture and segmentation," Marine and Petroleum Geology, vol. 28, pp. 1676–1689, 2011.

8. F. A. Stoakes, C. V. Campbell, R. Cass, and N. Ucha, "Seismic stratigraphic analysis of the Punta del Este Basin, offshore Uruguay, South America," American Association of Petroleum Geologists Bulletin, vol. 75, no. 2, pp. 219–240, 1991.

9. H. de Santa Ana, N. Ucha, and G. Veroslavsky, "Geología y potencial hidrocarburífero de las cuencas offshore de Uruguay," in Proceedings of the 5th Seminario Internacional: Exploración y Producción de Petróleo y Gas, Lima, Peru, 2005.

10. H. de Santa Ana, G. Veroslavsky, and E. Morales, "Potencial exploratorio del offshore de Uruguay,"Revista de la Industria Petrolera, no. 12, pp. 48–59, 2009.

11. D. A. Giberto, C. S. Bremec, E. M. Acha, and H. Mianzan, "Large-scale spatial patterns of benthic assemblages in the SW Atlantic: the Río de la Plata estuary and adjacent shelf waters," Estuarine, Coastal and Shelf Science, vol. 61, no. 1, pp. 1–13, 2004.

12. R. G. Peterson and L. Stramma, "Upper-level circulation in the South Atlantic Ocean," Progress in Oceanography, vol. 26, no. 1, pp. 1–73, 1991.

13. F. J. Hernández-Molina, M. Paterlini, L. Somoza et al., "Giant mounded drifts in the Argentine Continental Margin: origins, and global implications for the history of thermohaline circulation,"Marine and Petroleum Geology, vol. 27, no. 7, pp. 1508–1530, 2010.

14. F. J. Hernández-Molina, M. Paterlini, R. Violante et al., "Contourite depositional system on the Argentine slope: an exceptional record of the influence of Antarctic water masses," Geology, vol. 37, no. 6, pp. 507–510, 2009.

15. A. R. Viana, "Chapter 23 Economic Relevance of Contourites," Developments in Sedimentology, vol. 60, pp. 491–510, 2008.

16. T. Saeki, T. Fujii, T. Inamori, et al., "Delineation of methane hydrate concentrated zone using 3D seismic data in the eastern Nankai Trough," in Proceedings of the 6th International Conference on Gas Hydrates (ICGH '08), Vancouver, Canada, July 2008.

17. H. de Santa Ana, N. Ucha, L. Gutiérrez, and G. Veroslavsky, "Gas hydrates: estimation of the gas potential from reflection seismic data in the continental shelf of Uruguay," Revista de la Sociedad Uruguaya de Geología, no. 11, pp. 46–52, 2004.

18. S. Neben, B. Schreckenberger, J. Adam, et al., Research Cuise BGR04, ARGURU—Geophysical Investigations Offshore Argentine and Uruguay with Akademik Alexandr Karpinsky. Buenos Aires-Buenos Aires, 19/11/ 2004 –19/12/2004. Cruise Report and Preliminary Results, Bundesanstalt für Geowissenschaften und Rohstoffe, Hannover, Germany, 2005.

19. C. Hensen, M. Zabel, K. Pfeifer et al., "Control of sulfate pore-water profiles by sedimentary events and the significance of anaerobic oxidation of methane for the burial of sulfur in marine sediments,"Geochimica et Cosmochimica Acta, vol. 67, no. 14, pp. 2631–2647, 2003.

20. S. A. Barclay, R. H. Worden, J. Parnell, D. L. Hall, and S. M. Sterner, "Assessment of fluid contacts and compartmentalization in Sandstone reservoirs using inclusions: an example from the magnus oil field, North Sea," AAPG Bulletin, vol. 84, no. 4, pp. 489–504, 2000.

21. G. F. Tavella and C. G. Wright, "Cuenca del Salado," in Geología y Recursos Naturales de la Plataforma Continental Argentina, V. A. Ramos and M. A. Turic, Eds., pp. 95–116, 1996, Relatorio del XIII° Congreso Geológico Argentino y III° Congreso de Exploración de Hidrocarburos, Buenos Aires.

22. H. de Santa Ana, G. Veroslavsky, and E. Morales, "Estado exploratorio de la región costa afuera de Uruguay," in Proceedings of the 7th Congreso de Exploración y Desarrollo de Hidrocarburos, pp. 649–657, Instituto Argentino del Petroleo y del Gas, Mar del Plata, Argentina, 2008.

23. P. Meldahl, R. Heggland, B. Bril, and P. de Groot, "Identifying faults and gas chimneys using multiattributes and neural networks," Leading Edge, vol. 20, no. 5, pp. 474–482, 2001.

24. J. Tomasini, H. de Santa Ana, and A. Johnson, "Identification of new seismic evidence regarding gas hydrate occurrence and gas migration pathways offshore Uruguay," in Proceedings of the AAPG 2010 Annual Convention & Exhibition, New Orleans, La, USA, 2010.

25. E. C. Capen, "Probabilistic reserves! Here at last?" SPE Reservoir Evaluation and Engineering, vol. 4, no. 5, pp. 387–394, 2001.

26. Z. Lu and N. Sultan, "Empirical expression for gas hydrate stability law, its volume fraction and mass-density at temperatures 273.15 K to 290.15 K," Geochemical Journal, vol. 42, no. 2, pp. 163–175, 2008.

27. J. Tomasini, H. de Santa Ana, P. Gristo, et al., "Determination of Hydrate Formation Gases for Marine Gas Hydrates Offshore Uruguay," Project FSE_2009_53, ANII. In preparation.

28. S. Oliveira, O. Vilhena, and E. da Costa, "Time-frequency spectral signature of Pelotas Basin deep water gas hydrates system," Marine Geophysical Researches, vol. 31, no. 1, pp. 89–97, 2010.

29. A. Sad, D. Silveira, S. Silva, R. Maciel, and M. Machado, "Marine gas hydrates along the Brazilian margin," in Proceedings of the

ABGP/AAPG International Conference and Exhibition, Rio de Janeiro, Brazil, 1998, AAPG Search and Discovery Article #90933.

30. J. S. Booth, M. M. Rowe, and K. M. Fischer, "Offshore gas hydrate sample database with an overview and preliminary analysis," Open-File Report 96-272, U.S. Geological Survey, 1996,http://pubs.usgs.gov/of/1996/of96-272/.

Chapter 6

Study of Knocking Effect in Compression Ignition Engine with Hydrogen as a Secondary Fuel

R. Sivabalakrishnan[1] and C. Jegadheesan[2]

[1]Department of Mechatronics, SNS College of Technology, Coimbatore, India

[2]Department of Mechatronics, Kongu Engineering College, Perundurai, Erode, India

ABSTRACT

The aim of this project is detecting knock during combustion of biodiesel-hydrogen fuel and also the knock is suppressed by timed injection of diethyl ether (DEE) with biodiesel-hydrogen fuel for different loads. Hydrogen fuel is an effective alternate fuel in making a pollution-free environment with higher efficiency. The usage of hydrogen in compression ignition engine leads to production of knocking or detonation because of its lower ignition energy, wider flammability range, and shorter quenching distance. Knocking combustion causes major engine damage, and also reduces the efficiency. The method uses the measurement and analysis of cylinder pressure signal for various loads. The pressure signal is to be converted into frequency domain that shows the accurate knocking combustion of fuel mixtures. The variation of pressure signal is gradually increased and smoothly reduced to minimum during normal combustion. The rapid rise of pressure signal has occurred during knocking combustion. The experimental setup was mainly available for evaluating the feasibility of normal combustion by comparing with the signals from both fuel mixtures in compression ignition engine. This method provides better results in predicting the knocking feature of biodiesel-hydrogen fuel and the usage of DEE provides complete combustion of fuels with higher performance, and lower emission.

INTRODUCTION

The demand for fossil fuels gets increased by more usage of transportation and automobile. The use of fossil fuels emits more emissions such as HC, CO, CO_2, and NO_x and also makes harmful environmental condition. The best solution for this problem is to move on to alternative fuels. Hydrogen is the most effective alternative fuel which reduces the emission and fuel consumption and also provides better performance. Hydrogen has some limitations such as backfire and preignition. Saravanan et al. [1] proposed that the direct injection (DI) diesel engine was used to test the performance and emission of an engine. Hydrogen was injected at the intake port of the engine and diesel can be used as an ignition source. In order to improve the efficiency, the knocking combustion occurred as a major problem due to some properties of hydrogen fuel such as wider flammability range and shorter quenching distance. The biodiesel can be used as an ignition source instead of diesel which reduces the emissions of particulate matter and limits the autoignition condition. There is a possible minimum emission of NO_x at higher load conditions.

Zhen et al. [2] projected that the knock detection is to be done on several types of methods. These methods are in-cylinder pressure analysis, heat transfer analysis, light radiation, cylinder block vibration analysis, intermediate radicals and species analysis, ion current analysis, and exhaust gas temperature analysis. The most suitable methods are in-cylinder pressure analysis and heat transfer analysis. The knock intensity is the maximum amplitude of cylinder pressure fluctuation and rapid increase of pressure signal and heat release rate provides the information about abnormal combustion.

Wannatong et al. [3] determined that the knocking in engines leads to damaging the engine and limits the performance of the engine. The combustion and knock characteristics can be determined for diesel and dual fuel (Diesel and Natural Gas) by varying the temperature of intake mixture, increasing the amount of natural gas, mixture of diesel and natural gas. Engine knocks were noted for every increase of temperature of intake mixture and increasing the amount of natural gas. In this process, the higher intake temperature fastened the combustion and made autoignition of fuel before flame arrival. The rapid increase of cylinder pressure has shown the onset of knock in engine.

The knock detection method is to be done on the cylinder pressure, block vibration, and sound pressure signal in spark ignited (SI) engine. The three knock harmonic frequencies were estimated by analyzing the cylinder pressure signal under various operating conditions in spark ignited (SI) engine. The filtered pressure signal can be used to predict knock intensity and also helps to remove background noise. The knock windows and knock frequencies were determined by Lee et al. [4].

Brunt et al. [5] have made a comparison of calculated peak pressures at crank angle resolution for constant speed and also found out the peak knock pressure for all cycles. The measurement and analysis of cylinder pressure is used to obtain accurate knocking combustion. The knock intensity is to be determined by the maximum variability of peak pressure and its filtered data.

FUNDAMENTALS

Hydrogen Fuel

Hydrogen has clean burning characteristics that provide an efficient operation in CI engine. Hydrogen can be used as a secondary fuel in an internal combustion engine. The hydrogen burning combines with oxygen to form water and no other combustion products (except for little amounts of NO_x). Hydrogen cannot be ignited by compression due to higher autoignition temperature (585°C) than diesel fuel (180°C). Biodiesel is used as an ignition source for hydrogen fuel during combustion of compression ignition engine (Table 1).

Table 1: Fuels properties

Properties	Biodiesel	Hydrogen	Diethyl ether
Chemical formula	—	H2	C2H5OC2H5
Auto Ignition Temperature (K)	535	858	433
Calorific value (MJ/kg)	38.5	119.9	33.9
Density (kg/m3)	885	0.0837	713
Viscosity at 15.5°C, centipoises	—	—	0.023

Knock Fundamentals

Due to presence of some constituents in the fuel used, the rate of oxidation becomes so great that the last portion of the fuel-air mixture gets ignited instantaneously, producing an explosive violence, known as knocking. The explosive ignition of fuel-air mixture before the propagating flame is increasing successive cylinder pressure oscillations. The well-examined external mixing of hydrogen with the intake of air causes backfire and knock, especially at higher engine loads. The abnormal combustion of hydrogen fuel in CI engine will produce an increased chemical heat release rate, which results in a rapid pressure rise and higher heat rejections. The maximum amplitude of pressure oscillation and analysis of exhaust temperature is a good indicator for severity of the knock.

EXPERIMENTAL SETUP

In this study, a single cylinder, four strokes, water cooled direct injection diesel engine was operated as dual fuel engine which uses hydrogen and biodiesel shown in Figure 1. The engine details are shown in Table 2. Hydrogen fuel is stored in a storage cylinder. A pressure regulator was used to regulate hydrogen passed to flame arrester through flow control valve and check valve. Check valve is used to pass hydrogen in forward direction alone and it can be closed if any gas returns from CI engine. Flame arrester can have 3/4thfiled water in an enclosed tank to restrict backfire to hydrogen cylinder during combustion. Hydrogen fuel is fed at the inlet manifold in diesel engine. DEE is to be fed at the inlet port before the hydrogen port is used. A pressure transducer was used to pick up peak pressure oscillation during the combustion of fuel. The pressure signal is acquired by PC data acquisition system.

Table 2: Engine specification

Name	Specification
Type	4-Stroke, Single Cylinder Diesel
Engine Make	Kirloskar
Power	5.2 kW
Speed	1500 rpm
Stroke	110 mm
Bore	87.5 mm
Capacity	661 cc

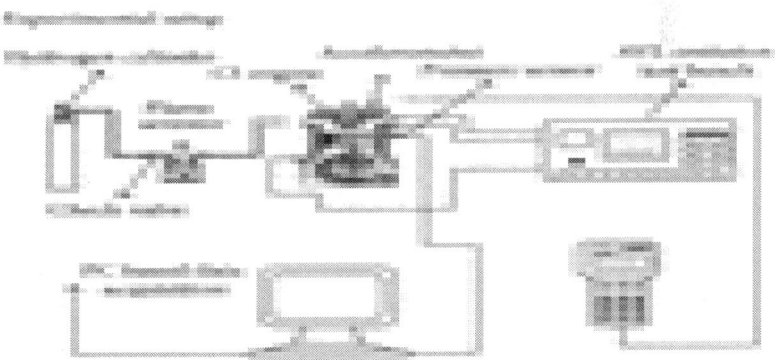

Figure 1: Experimental setup.

FREQUENCY ANALYSIS OF PRESSURE SIGNAL

The pressure transducer is used to record the in-cylinder pressure signal with respect to crank angle. This signal can be acquired using PC data acquisition system and the crank angle is got from rotary encoder coupled with crank shaft.

This signal is given to power spectral analysis tool in Lab View software which converts the given signal into frequency domain. The conversion of pressure signal into frequency domain is shown in Figure 2. The frequency signal is used to predict the knocking combustion of engine during abnormal conditions.

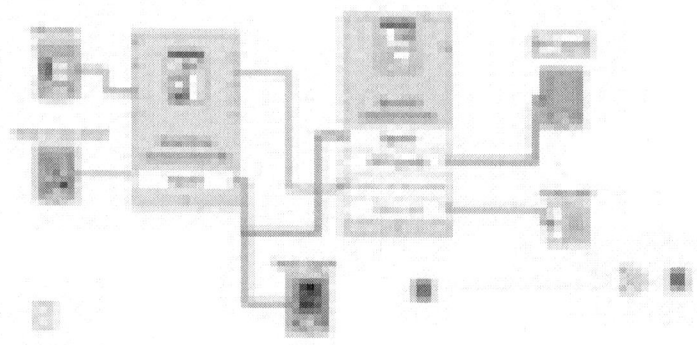

Figure 2: Program for FFT conversion.

RESULT AND DISCUSSION

Experimental tests were carried out for biodiesel-hydrogen mixtures and biodiesel-hydrogen mixture with DEE at various loads. The pressure signal variation and its power spectrum can be shown in Figures 3 and 4. The engine has been run on biodiesel-hydrogen mixtures from no load to full load. In normal combustion, the pressure signal gradually reaches the peak value after Top Dead Centre of the piston (TDC of greater than 3600 of crank angle) and again smoothly decreases to minimum value of the pressure.

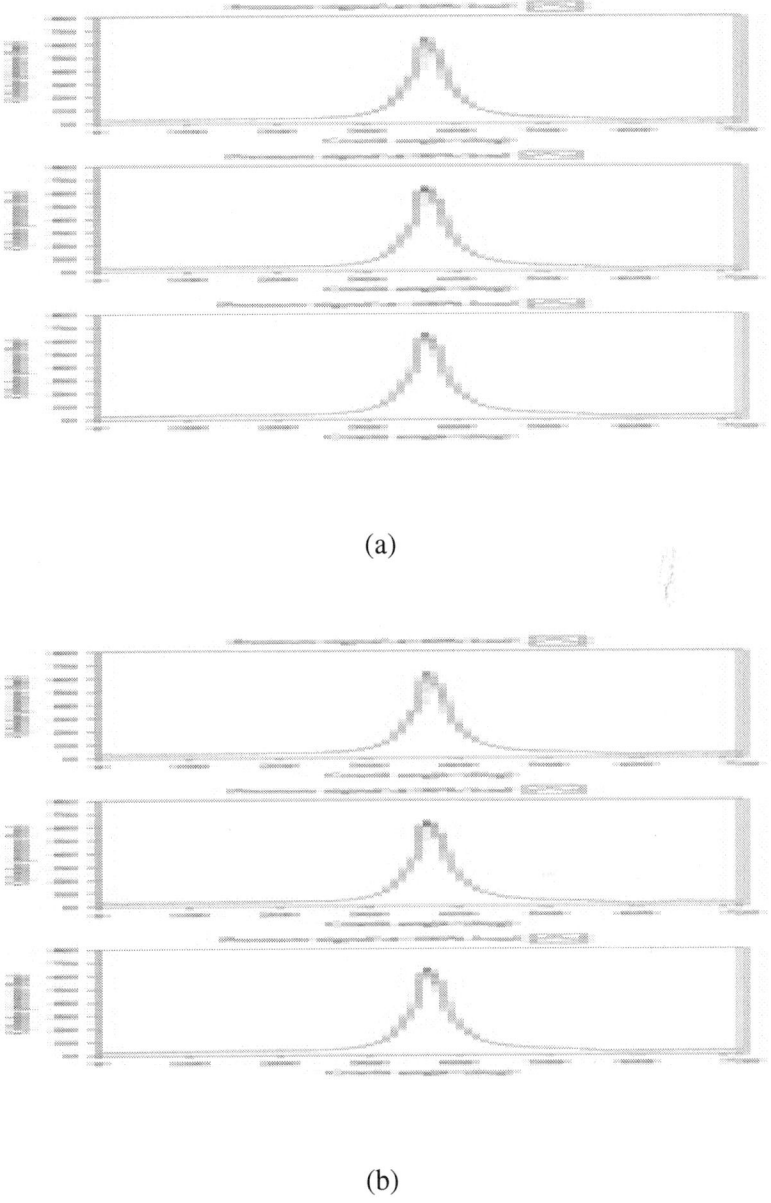

(a)

(b)

Figure 3: (a) In-cylinder pressure signal for biodiesel and hydrogen at various loads. (b) Power spectrum of pressure sign.

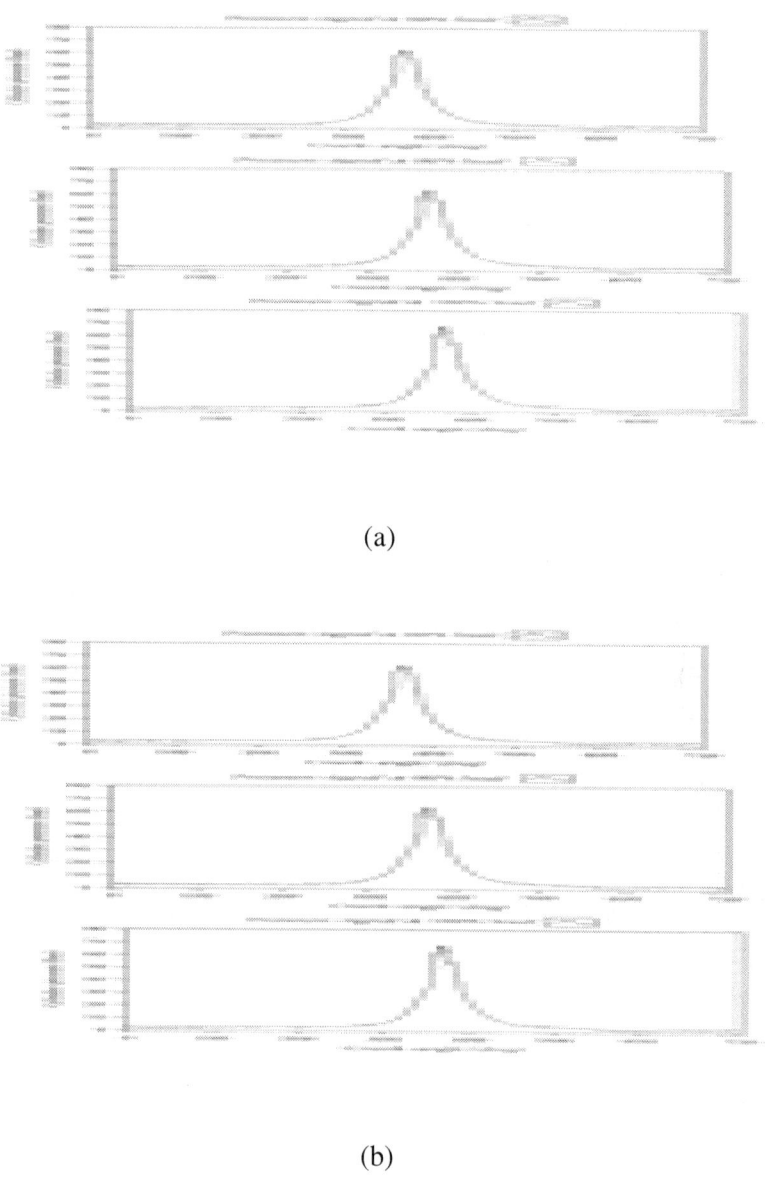

(a)

(b)

Figure 4: (a) In-cylinder pressure signal for biodiesel and hydrogen with DEE at various loads. (b) Power spectrum of pressure signal.

In knocking combustion, the peak pressure signal gets rapid oscillation at every crank angle. After crossing the load of 52%, there could be a maximum oscillation in peak pressure compared to light load, as well as a significant notification from power spectrum of pressure signal. From the power spectrum signal, the first harmonic knocking frequency can be found as 1.65 kHz for 70% and 80% load and second harmonic frequency is 2.4 kHz and 2.3 kHz for 70% and 80% load, respectively. There are no harmonic frequencies found for biodiesel-hydrogen with diethyl ether. Next, the engine was run on biodiesel-hydrogen mixture and diethyl ether can be injected at the intake valve opening moment in an engine. The different types of load can be applied to these mixtures and the signal can be noted down. This result shows that complete combustion of engine during the application of higher loads. Along with the analysis of pressure signal, the exhaust gas temperature and brake specific fuel consumption can be considered to find out the knocking behavior of the engine.

Combustion Characteristics

The cylinder peak pressure variation and its power spectrum are given in Figures 3 and 4. The peak pressure and pressure oscillation are higher for the biodiesel-hydrogen fuel mixture when compared to the diethyl ether. The biodiesel fuel can act as a main fuel which can be injected at direct injection port and hydrogen is supplied at intake manifold whose flow rate is fixed at 0.5 lpm. In biodiesel-hydrogen, the hydrogen fuel properties make the abnormal combustion in compression ignition engine. This can be got from analysis of pressure signal and its power spectrum. The pressure signal can be got from PC data acquisition system which is given in the LabView software. This can be converted into frequency domain. In part load, there is no rapid rise or oscillation of pressure signal during combustion phase. This shows that the complete combustion fuel mixture takes place at minimum load. After injecting the diethyl ether with the biodiesel-hydrogen, there are no changes in pressure signal during minimum (<60%) load. The flow rate of diethyl ether is optimized at 0.25 g/min, according to the signal got from the engine during the operation. The diethyl ether helps to reduce the abnormal combustion to take place at maximum (>60%) load. The diethyl ether reduces the peak pressure occurring during the combustion of fuel due to lag in ignition timing and acts as an ignition improver. The autoignition can be prevented by supplying diethyl ether as an additive. The knocking combustion can be found at higher load and after applying diethyl ether smooth combustion of fuel mixture takes place inside the engine.

Performance Characteristics

The performance of biodiesel-hydrogen fuel and biodiesel-hydrogen fuel with DEE can be shown in Tables 3and 4, respectively. The performance can be noted for various applications of load up to 80% load. The exhaust temperature is taken from the thermocouple sensor. The performance of engine during knocking and nonknocking can be evaluated using these equations.

Table 3: Performance of biodiesel and hydrogen

Sl. no.	Load (kg)	Exhaust temerature T3(°C)	Indicated power, IP (kW)	Brake power, BP (kW)	BSFC = FC/BP (kg/kW-hr)	Mech Efficiency = BP/IP (%)
1	0	198	2.68715	0.2835	1.84282	10.55151
2	2	229	3.084022	0.8506	0.69387	27.58103
3	4	247	3.472626	1.4176	0.46834	40.8243
4	6	273	3.753743	1.9847	0.37341	52.87376
5	8	338	4.274637	2.5518	0.32021	59.69665
6	10	375	4.812067	3.1188	0.30051	64.81384
7	12	410	5.275084	3.6859	0.28817	69.81384
8	14	505	5.820782	4.2530	0.27749	73.06622
9	16	564	6.449162	4.8200	0.27545	74.73987

K W: Killowatt, °C: Degree Celcius, kg: killogram; hr: hourhenry.

Table 4: Performance of biodiesel and hydrogen with DEE.

Sl. no.	Load (kg)	Exhaust temerature T3 (°C)	Indicated power, IP (kW)	Brake power, BP (kW)	BSFC = FC/BP (kg/kW-hr)	Mech Efficiency = BP/IP (%)
10	0	194	2.48045	0.283	2.14154	11.4308
11	2	224	2.70369	0.850	0.80882	31.46093

12	4	255	3.47263	1.417	0.55646	40.8243
13	6	291	3.81162	1.984	0.42659	52.07091
14	8	327	4.23330	2.551	0.38290	60.27963
15	10	361	4.72112	3.118	0.34491	66.06244
16	12	404	5.21721	3.685	0.30774	70.64998
17	14	451	5.69676	4.253	0.30015	74.65692
18	16	520	6.18458	6.184	0.29316	77.9373

K W: Killowatt, °C: Degree Celcius, kg: killogram; hr: hourhenry.

The power and efficiency can be calculated from these formulas.

- Indicated power

$$IP = \frac{nP_{mi}LANk * 10}{6} \text{ kW,}$$

(1)

where P_{mi} indicated mean effective pressure in bar, n indicated number of cylinders, L indicated length of stroke in m, A indicated area of piston in m^2, N indicated speed in rpm, and K indicated ½ (for four-stroke engine).

- Brake power

$$BP = \frac{2\pi NT}{60 * 1000} \text{ kW,}$$

(2)

Where N is speed in rpm and T is torque in Nm.

- Mechanical efficiency

$$\eta_{mech} = \frac{BP}{IP}.$$

(3)

Figure 5 shows the variation of exhaust gas temperature with respect to load. It is observed that the exhaust gas temperature of biodiesel-hydrogen is similar to that of those fuel mixtures along with DEE for below 60% of load. When the amount of load was increased, the engine experienced knocking level due to improper combustion of fuel (fuel mixture remains same to find out knocking level). The exhaust gas temperature gets increased for the load above 70% due to late combustion of fuel increasing the exhaust gas temperature. The hydrogen fuel gets accumulated in full throttle running of an engine during higher load. The injection of diethyl ether leads to providing normal combustion of engine, and the complete combustion of fuel takes place due to timed injection of DEE at the inlet port.

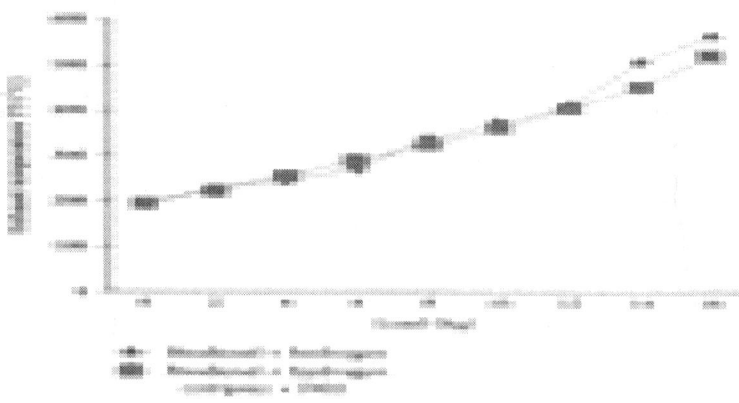

Figure 5: Exhaust temperature variation with load.

Figure 6 shows the variation of brake specific fuel consumption for various fuel mixtures with respect to load. The brake specific fuel consumption is mainly based on the torque delivered by the engine with respect to the mass flow rate of fuel delivered to the engine.

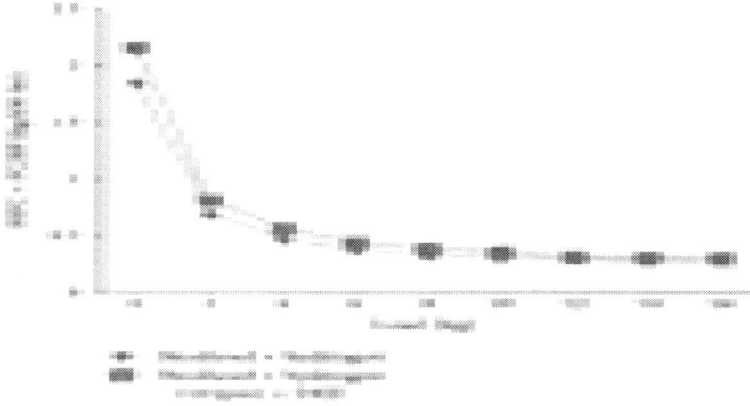

Figure 6: Brake specific fuel consumption variation with load.

It is observed that the brake specific fuel consumption of biodiesel-hydrogen with DEE is decreased with the load increasing to maximum. In case of hydrogen-biodiesel, brake specific fuel consumption is increased because of knocking combustion. When there is a decrease in brake specific fuel consumption, it also decreases the brake thermal efficiency of the engine. The brake specific fuel consumption is well decreased at minimum load compared to higher load, while applying diethyl ether during the combustion of fuel mixture.

Figure 7 shows the variation of mechanical efficiency for various fuel mixtures with respect to load. The mechanical efficiency is defined as the ratio of brake power to the indicated power. It is observed that the mechanical efficiency of biodiesel-hydrogen with DEE increases for load above 50%. There is a slight increase of mechanical efficiency for the 10% load. The increase in mechanical efficiency in the case of hydrogen-biodiesel with DEE operation is mainly due to higher charge intake leading to complete combustion and the energy release is higher in case of DEE. The diethyl ether helps to make complete burning of fuel during combustion at higher load.

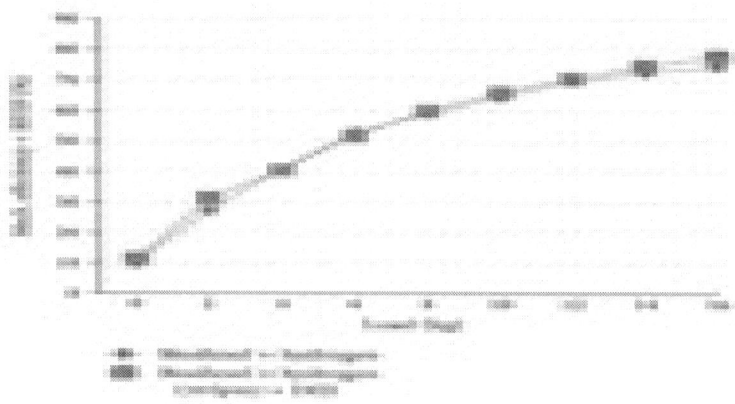

Figure 7: Mechanical efficiency variation with load.

CONCLUSIONS

An experimental model of knock detection for biodiesel-hydrogen fuel and biodiesel-hydrogen fuel mixtures with diethyl ether has been developed. The most suitable knock techniques have been applied to detect knock in compression ignition engine.

- The knock measurement and analysis can be done for the biodiesel-hydrogen fuel and biodiesel-hydrogen fuel with DEE.
- The pressure signal could be got from a pressure transducer and converted into frequency domain for analysis of the knock.
- The exhaust temperature can also be used to find out the knocking combustion for the same fuel mixture (biodiesel-hydrogen fuel at 10 lpm) at higher loads.
- The performance and knock limiting operation of engine could be improved by using DEE as an additive fuel.
- The diethyl ether is taken to suppress the knocking behaviour in compression ignition engine during combustion of mixture of hydrogen-biodiesel fuel.

The performance characteristics of both hydrogen-biodiesel fuel and hydrogen-biodiesel fuel with DEE could be computed for various applications of load.

REFERENCES

1. N. Saravanan, G. Nagarajan, C. Dhanasekaran, and K. M. Kalaiselvan, "Experimental invetigation of hydrogen fuel injection in DI dual fuel diesel engine," SAE Paper 2007-01-1465, 2007.

2. X. Zhen, Y. Wang, S. Xu et al., "The engine knock analysis—an overview," Applied Energy, vol. 92, pp. 628–636, 2012.

3. K. Wannatong, N. Akarapanyavit, and S. Siengsanorh, "Combustion and knock characteristics of natural gas diesel dual fuel engine," JSAE 2007-01-2047, 2007.

4. J.-H. Lee, S.-H. Hwang, J.-S. Lim, D.-C. Jeon, and Y.-S. Cho, "New knock-detection method using cylinder pressure, block vibration and sound pressure signals from a SI engine," in Proceedings of the SAE International Spring Fuels & Lubricants Meeting & Exposition, pp. 27–38, May 1998.

5. M. F. J. Brunt, C. R. Pond, and J. Biundo, "Gasoline engine knock analysis using cylinder pressure data," in Proceedings of the SAE International Congress & Exposition, pp. 21–33, February 1998.

Chapter 7

Co$_2$ Capture in Ionic Liquids: A Review of Solubilities and Experimental Methods

Elena Torralba-Calleja, James Skinner, and David Gutiérrez-Tauste

Renewable Energies R&D Department, LEITAT Technological Center, Carrer de la Innovació, 2, Terrassa 08225 Barcelona, Spain

ABSTRACT

The growing concern of climate change and global warming has in turn given rise to a thriving research field dedicated to finding solutions. One particular area which has received considerable attention is the lowering of carbon dioxide emissions from large-scale sources, that is, fossil fuel power. This paper focuses on ionic liquids being used as novel media for CO_2 capture. In particular, solubility data and experimental techniques are used at a laboratory scale. Cited CO_2 absorption data for imidazolium-, pyrrolidinium-, pyridinium-, quaternary-ammonium-, and tetra-alkyl-phosphonium-based ionic liquids is reviewed, expressed as mole fractions (X) of CO_2 to ionic liquid. The following experimental techniques are featured: gravimetric analysis, the pressure drop method, and the view-cell method.

INTRODUCTION

In recent years, increasing attention has been paid towards the worldwide climate change. Moreover, the exponential increase of carbon dioxide emissions into the atmosphere from the combustion of fossil fuels, making up the 86% of greenhouse gases [1], does not reflect a sustainable energy model. Entry into the Kyoto protocol has brought about the need to reduce anthropogenic emissions of CO_2. Thus carbon capture and storage (CCS) proves to be one of the most important initiatives to mitigate this global warming effect.

CCS is a concept based on the reduction of CO_2 emissions into the atmosphere from industrial processes, such as ammonia production, natural gas processing, or cement manufacture, to name a few. This review however will focus on CO_2 emissions from fossil fuel power plants, which is seen to be the main contributor to this effect [2]. It has been approximated that, if CCS is fully implemented, its potential by 2050 could be the total capture and storage of 236 billion tons of CO_2 [3]. An approach to CCS that holds the greatest promise is the sequestration of captured carbon dioxide, in suitable deep sedimentary formations, for example, oil and depleted gas reservoirs, coal beds, and saline deposits [4–7]. The challenge is to develop a technology which will allow us to accomplish this task in an environmental, economic, and efficient way in the next years [8–10]. However the need to assess the environmental impact is great. The potential risks of geological storage to humans and ecosystems are abundant and need to be carefully monitored. Leakage of sequestered CO_2 would be the main concern. This could happen along fault lines, ineffective confining layers, abandoned wells, and so forth. The pollution of groundwater and mineral deposits is also a problem and could have lethal effects on plant life and animals. A recent review by Manchao et al. [11] offers a detailed risk assessment of the CO_2 injection process and storage in geological formations, with a main focus on abandoned coal mines and coal seams.

An alternative to geological storage of CO_2 would be the direct conversion of CO_2 into a high-valued product after the initial capture; this is sometimes referred to as carbon capture and usage (CCU). CO_2 is used in many industries such as the food industry (carbonation of beverages), electronics industry (surface cleaning and semiconductor manufacture), and the chemical industry (polymers, plastics, and fertilizers). CCU is yet to be a mainstream technology so that many process aspects and methods are being published and reviewed [12, 13].

STATE-OF-THE-ART CO$_2$ CAPTURE TECHNOLOGIES

The capture of CO_2 is achieved through the use of specific materials that interact with the gas in one form or another. The materials that are used depend on the processes in which the flue gas is conditioned (Figure 1) [14].

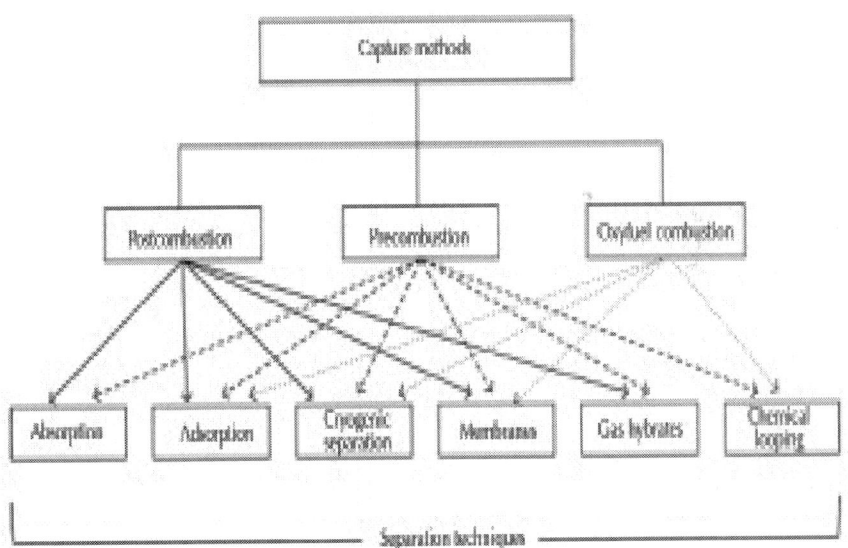

Figure 1: Possible techniques that can be used in conjunction with the processes of postcombustion, precombustion, and oxyfuel combustion.

There are three processes, each of which conditions the CO_2 for capture in different ways.

Postcombustion. The separation of CO_2 from the flue gas after the combustion of fuel. Air is typically used as the oxidant in this process; therefore the flue gas becomes largely diluted with nitrogen.

Precombustion. The hydrocarbon fuel (in this case gasified coal) is converted into carbon monoxide (CO) and hydrogen (H_2). This forms a synthesis gas. By using water shift conversion, CO is converted into CO_2. Finally the CO_2 is then separated from the H_2.

Oxyfuel CO_2 Combustion. It uses pure oxygen as the oxidant instead of air, creating a flue gas mainly consisting of high-concentrated CO_2 and steam.

Although CO_2 capture and separation is a well-known technology, this technology is just applied in a small scale, so that right now it is not commercially available for being used in large power stations. The most challenging obstacle to overcome in CCS and CCU is finding an effective technique that satisfies environmental and economic factors. Some of the currently studied techniques for capturing CO_2 from the three conditioning processes are as follows (Figure 1).

Absorption occurs within the bulk of the material via a chemical or physical interaction. Chemical absorbents react with the CO_2, forming covalent bonds between the molecules. The solvent can be habitually regenerated through heating and captured CO_2 is released. This mechanism can also be made highly selective by the introduction of specific chemical complexes. Typical compounds used in this process are amines, or ammonia-based solutions. Physical absorbents obey Henry's law, where gas solubility is directly proportional to the partial pressure of the said gas in equilibrium, at a constant temperature. Typically this is at high CO_2 partial pressures and low temperatures. The interaction between CO_2 and the solvent is by nonchemical surface forces, that is, Van der Waals interaction. Regeneration of the solvent is achieved by increasing the temperature and lowering the pressure of the system [15]. Selexol and Rectisol are examples of physical absorbents that have been used in natural gas sweetening and synthesis gas treatment.

Adsorption, as opposed to absorption, takes place at the surface of the material. This interaction can also occur chemically (covalent bonding) or physically (Van der Waals). Typical adsorbers are solid materials with large surface areas, such as zeolites, activated carbons, metal oxides, silica gel, and ion-exchange resins. These can be used to capture CO_2 by separation, so that flue gas is put in contact with a bed of these adsorbers, allowing the CO_2 capture from the other gases which pass through. When the bed is fully saturated with CO_2, the flue gas is directed to a clean bed and the saturated bed is regenerated [16]. Three techniques can be employed to the adsorption mechanism: pressure swing adsorption (PSA) introduces the flue gas at high pressure until the concentration of CO_2 reaches equilibrium, then the pressure is lowered to regenerate the adsorbent, temperature swing adsorption (TSA) increases the temperature to regenerate the adsorbent, and electric swing adsorption (ESA) is where a low-voltage electric current is passed through the sorbent to regenerate. Adsorption is not yet considered practical for large-scale applications as the CO_2 selectivity in current sorbents is low. However, recently new sorbents are being investigated such as metal-organic frameworks and functionalised fibrous matrices that show some promise for the future of this particular technique.

Membrane separation technology is based on the interaction of specific gases with the membrane material by a physical or chemical interaction. Through modifying the material, the rate at which the gases pass through can be controlled. There are wide varieties of membranes available for gas separation, including polymeric membranes, zeolites, and porous inorganic membranes, some of which are used in an industrial scale and have the possibility of being implemented into the process of CO_2 capture. However achieving high degrees of CO_2 separation in one single stage has so far proved to be difficult; therefore, having to rely on multiple stages has led to increasing energy consumption and cost. An alternative approach is to use porous membranes as platforms for absorption and stripping. Here a liquid (typically aqueous amine solutions) provides the selectivity towards the gases. As the flue gas moves through the membrane, the liquid selects and captures the CO_2 [17].

Cryogenic Separation is a technique based on cooling and condensation. This has the advantage of enabling the direct production of liquid CO_2, benefiting transportation options. Although a major disadvantage of cryogenic technology in this respect is the high amounts of energy required to provide cooling for the process, this is especially prominent in low-concentration gas streams [18]. This technique is more suited to high-concentration and high-pressure gases, such as in oxyfuel combustion and precombustion.

Within these techniques lie the materials with which research pathways aim to develop more effective CO_2 capture mechanisms. Currently the postcombustion process is the most widely researched area for reducing CO_2 emissions from power stations. This is mainly because it can be retrofitted to existing combustion systems without a great deal of modification, unlike the other two processes. The flue gas emitted, from the postcombustion of fossil fuels in power stations, has a total pressure of 1-2 bars with a CO_2 concentration of approximately 15%. As this process creates low CO_2 concentration and partial pressures, strong solvents have to be used to capture the CO_2, resulting in a large energy input to regenerate the solvent for further use. This creates the technical challenge of finding an efficient, costeffective, and low-energy-demanding capture mechanism using novel materials.

Aqueous Amines Used in Postcombustion

The conventional technologies used in this postcombustion process are solvent-based chemical absorbers. The common chemical solvents used for separation are aqueous amines, which are ammonia derivatives, where one or more of the hydrogen atoms have been replaced by alkyl groups. Some common amines used in this process are (Table 1) monoethanolamine (MEA) [19], methyldiethanolamine (MDEA) [20], and diethanolamine (DEA) [21]. Aqueous amines are stated as "conventional absorbers" because they are well-known solvents used in the oil and gas industries, dating back to the 1930s; for example, Gregory and Scharmann investigated the implementation of amine CO_2 scrubbers in a hydrogenation plant of the Standard Oil Company of Louisiana in 1937. Today the aqueous amine absorption technology is still used in natural gas sweetening (removal of acidic gases, for example, hydrogen sulphide and carbon dioxide) and has also been applied to some small-scale fossil fuel power plants [22, 23], for example, Fundación Ciudad de la Energía (CIUDEN), Alstom power plant, and so forth.

Table 1: Chemical structures of commonly used amines

Amine	Acronym	Structure
Monoethanolamine	MEA	$H_2N-CH(H)-CH(OH)(H)-H$
Methyldiethanolamine	MDEA	$HO-H_2C-H_2C-N(CH_3)-CH_2-CH_2-OH$
Diethanolamine	DEA	$HO-H_2C-H_2C-N(H)-CH_2-CH_2-OH$
Piperazine	PIPA	Piperazine ring structure

Briefly, post-combustion capture with amines, seen in Figure 2, involves the CO_2 being removed by circulating a flue gas stream into a chamber containing an aqueous amine solution. In the case of primary amines like MEA, the CO_2 is captured by a chemical absorption process in which the CO_2 reacts with the amine in the form of a carbamate [24]. With secondary and tertiary amines, which do not possess a hydrogen atom attached to a nitrogen atom, they react with CO_2 in the form of bicarbonate through hydrolysis. This is a reversible reaction, and at high temperatures the captured CO_2 is released and the amine solution recycled. Piperazine (PIPA) is commonly used to improve reaction kinetics of secondary and tertiary amines in the form of an additive; this is because the heat of reaction to form a bicarbonate is low, causing more heat being needed for regeneration and thus higher costs [25].

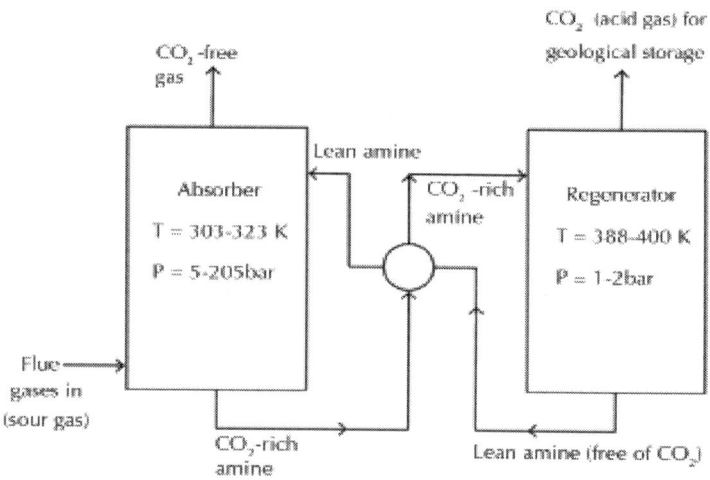

Figure 2: Conceptual scheme for CO_2 absorption using amine-based chemical absorption.

Amines are so effective for CO_2 capture thanks to some of their properties such as high reactivity with CO_2, high absorbing capacity (in terms of mass of CO_2), relatively high thermal stability, and CO_2 selectivity [26]. However there are inherent disadvantages linked with amines, which need to be addressed in order to make a valid and efficient process for CO_2 capture. These disadvantages come in the form of high vapour pressure, corrosive nature, and high-energy input for regeneration. The high vapour pressure allows emission of amine gases into the air upon heating. These gases are unstable in nature thus giving them the possibility of producing dangerous toxins such as nitrosamines, nitramines, and amides. Nitrosamines are of the most concern as they are carcinogenic and toxic to humans even at low levels [27]. Amines are also corrosive, especially MEA. They take part in reactions in which waste forms and can eventually corrode the equipment, Kittel et al. [28] investigated the effects of MEA operating pilot plants and found that areas made of carbon steel had corrosion rates of 1 mm year^{-1}; so besides environmental impacts, expense on a large industrial scale is another issue. The recycling/regeneration process leads to high-energy consumption in order to break the chemical bonds formed between the CO_2 and amine [29]. This process also causes degradation of the amine which limits its CO_2 capture rate, causing them to be replaced frequently.

Much research has gone into developing new solvents with the foresight of being superior to amines. The factors that would allow new solvents to perform better than amines are lower cost, lower volatility, better thermal stability, less degradation, low corrosive nature, and low energy needed for regeneration and adaptability to an existing system. Although amines have high CO_2 solubility and selectivity, environmental and economic effects are taken into consideration when selecting the criteria for the most suited CO_2 capture mechanism. While continued research into improving the performance of these mature technologies is expected, research into novel materials and technologies could produce the significant breakthroughs required to minimise the environmental and energy penalties of capture.

Ionic Liquid Media for Co$_2$ Capture

One of these advanced R&D pathways currently conveying great potential in the field of alternative technologies is ionic liquids (ILs). ILs are commonly defined as materials that are comprised of large organic cations and organic/inorganic anions, which demonstrate melting points below 100°C [30]. To date a wide range of ILs has been synthesised through different combinations of anions and cations. It has been stated that the theoretical number of potential ILs is to the order of 10^{18} [31]. An example of some of the common cations and anions used in IL synthesis can be seen in Table 2.

Table 2: Structures of common IL cations and anions

Cation	Structure	Anion	Structure
Imidazolium		Tetrafluoroborate	
Pyrrolidinium		Hexafluoroborate	
Pyridinium		Bis(trifluorophoshate) imide	
Quaternary Ammonium		Nitrate	
Tetra Alkyl Phosphonium		Acetate	

ILs possess several unique and diverse characteristics such as high thermal and chemical stability, low vapour pressure, large electrochemical window, tuneable/designer nature, and excellent solvent properties for a range of polar and nonpolar compounds. It is due to these characteristics that research into developing and implementing ILs over the past decade has spanned into many sectors of industry [32], for applications such as electrolytes [33], solar cells [34], lubricants [35], electropolishing and electroplating [36], and biomass processing [37], to name a few. This has become possible due to the large number of ILs that can be synthesised in the lab [38–40] and purchased commercially. Companies like BASF, Merck, Sigma-Aldrich, Solvionic, Sachem, and IoLitec provide basic ILs and can also aid in the design and development of ILs for specific tasks. Therefore these compounds have created exciting new media for emerging technological applications already commercially available.

Ionic Liquids in the Scope of Co₂ Capture

The aforementioned characteristics are particularly advantageous when applying ILs as solvents for CO_2 capture in comparison to current aqueous amine technology:

- less energy is required when regenerating ILs to remove the captured CO_2 [41] due to their physical absorption mechanism,
- further efficiency is attained by their low vapour pressure, which allows them to be regenerated and reused with no appreciable losses into the gas stream [42, 43],
- ILs have a high thermal and chemical stability; typically they degrade at temperatures >300°C [44] avoiding their reaction with impurities and causing corrosion to the equipment,
- the tuneable and designer nature of ILs offers many options concerning the physicochemical properties (viscosities and densities [45–47], heat capacities [48], thermal decomposition temperatures [49], surface tension [50], toxicity and health issues [51, 52], and corrosion [53, 54]) in the sense that the anions and cations can be manipulated to create an IL for a specific task.

This designer aspect can also be applied to the anion or cation in the sense that various chemical functionalities and structures can be attached, allowing properties such as absorption and viscosity to be controlled. These are commonly referred to as task-specific ionic liquids (TSILs). Generally ILs fulfil many of the major requirements stated in the green-chemistry principles stated by Anastas and Warner [55], in that they offer a new approach to industrial/chemical processes whereby steps are taken to eliminate hazardous waste in a system before a by-product is formed, thus neglecting the use of volatile organic solvent.

Most development concerning ILs for CO_2 capture is at present conducted at laboratory scale, while other technological applications are already in use as it was mentioned before. Conversely, their industrial application and implementation is being constantly investigated in areas of post-combustion [56]. For industrial-scale integration, it is necessary to achieve extensive knowledge of their physical and chemical properties. Therefore the need for experimental techniques and data is critical in enabling the ionic liquid to be the green, viable, and economic carbon capture technique of the future.

The solubility of CO_2 in ILs compared to other gases such as methane and nitrogen enables ILs to separate CO_2 from the source, be it a power plants' flue gas or natural gas. Even when there are low concentrations of CO_2 in a mixed gas, the IL can be designed to incorporate a functional group, such as an amine, thus rendering it task specific. The capacity for CO_2 solubility in ILs originates from the asymmetrical combination of the anion and cation, which results from short-range repulsive forces between their ionic shells. Therefore the more incompatible the ionic constituents are the greater the solubility is.

Conventional Ionic Liquids

Over the past decade, and at present, research has been built upon measuring the effects of variables such as pressure, temperature, and anion/cation choice. Results have shown high carbon dioxide solubility in what have become known as conventional ionic liquids. They are defined as ILs that do not possess an attached functional group and have been reported by many as portraying the typical behaviour of physical solvents [57–59]. This is evident when low-pressure CO_2 (1-2 bars) is put in contact with the IL, resulting in low CO_2 concentrations in the liquid phase. As the increment of pressure increases, typically to up to 100 bar, the concentration of absorbed CO_2 increases. Thus displaying the general characteristics of a physical absorber. As a rule, the solubility of CO_2 in ILs increases with increasing pressure and decreases with increasing temperature. The physical absorption mechanism is a result of the interaction between the CO_2 molecules and the IL, in which the CO_2 occupies the "free space" within the ILs structure through a large quadrupole moment and Van der Waals forces.

Anion and Cation Effects

In order to create an optimal process for capturing CO_2 in ILs, assessment of the essential building blocks, that is, cation/anion combinations, needs to be investigated. Synthesising ILs that encompass CO_2-philic groups on the anion such as carbonyls or fluorines has proven to increase CO_2 capture [60]. In the past decade studies have shown that the origin of high solubility is strongly dependent on the choice of anion [61]. Aki et al. [62] investigated the influence of the anion with seven ILs. They all contained the 1-butyl-3-methylimidazolium [Bmim] cation. The results are shown in Table 3.

Table 3: Influence of anions in different ionic liquids

Anion	Nomenclature	Classification	Solubility of CO_2 in IL
Dicyanamide	$[DCA]^-$	Nonfluorinated anions	Low
Nitrate	$[NO_3]^-$		
Tetrafluoroborate	$[BF_4]$	Fluorinated anions	Relatively high
Hexafluorophosphate	$[PF_6]^-$		
Trifuoromethanesulfonate	$[TfO]^-$		
Bis(trifluoromethylsulfonyl) imide	$[Tf_2N]^-$		
Tris(trifluoromethylsulfonyl) methide	$[Methide]^-$		

Aki and coworkers also systematically investigated the effects of the cation on CO_2 solubility; they found that, in general, the increase of the alkyl chain on the cation resulted in a slight increase in solubility, which became more apparent at higher pressures. The effect of increasing the alkyl chain results in the increased volume available for CO_2 interaction. Muldoon et al. [60] concluded that adding partially fluorinated alkyl chains on the imidazolium cation does increase CO_2 solubility. They compared [hmim][Tf$_2$N] directly to [C$_6$H$_4$F$_9$mim][Tf$_2$N] and found that this increased solubility was due to fluorinating the last four carbons of the alkyl chain. Research on IL CO_2 solubility, in general, has focused intensively on imidazolium-based structures. However some groups have focused on using different cations. Recently Carvalho et al. [63] reported CO_2 solubilities in two phosphonium-based ILs, [THTDP] [Tf$_2$N] and [THTDP] [Cl]. They found exceptionally high solubility measurements exceeding those of current imidazolium-based ILs; they go on to conclude that their study shows the highest recorded solubility observed without chemical interactions in the absorption process. Although imidazolium is the most stable and commercially available cation of choice, it is evident that there are further enhancements and possibilities that can be developed from other bases.

To provide further insight into the interactions between CO_2 and the constituent anions and cations of RTILs, researches using spectroscopic approaches and molecular simulations have been made. Of which has broadened our understanding of absorption mechanisms and structure-property relationships, Kazarian et al. [64] used ATR-FTIR spectroscopy to analyse the specific interactions of CO_2 and ILs [Bmim][BF_4] and [Bmim][PF_6]. They saw evidence of chemical interactions between the anion [PF_6]$^-$ and CO_2. They concluded that they observed weak Lewis acid-base interactions, where the anion acts as a Lewis base. ILs by their nature have intrinsic acid-base properties. These properties can be enhanced with the addition of acidic functions like carbonic or halide acids; likewise, basic functions like amino and fluorine groups can be added. This has shown to create specific Lewis acid-base chemical interactions between CO_2 and the IL.

As it can be seen in Tables 4 and 5, fluorination of the anion and in some cases the cation can improve CO_2 solubility in RTILs. However the associated disadvantages are cost increase, poor degradability, and a negative environmental impact [65]. Therefore paths to develop ILs with enhanced CO_2 solubility without fluorination are also being investigated.

Table 4: CO_2 solubility data for imidazolium-based ionic liquids

Ionic liquid	Acronym	T(K)	P(bar)	xCO_2	References
1-N-Octyl-3-methylimidazolium hexafluorophosphate	$C_8mim[PF_6]$	313	92.67	0.7550	Blanchard et al. 2001 [86]
1-N-Butyl-3-nethylimidazolium nitrate	$Bmim[NO_3]$	323	92.62	0.5300	Blanchard et al. 2001 [86]
1-N-Octyl-3-methylimidazolium tetrafluoroborate	$C_8mim[BF_4]$	313	92.90	0.7080	Blanchard et al. 2001 [86]
1-Ethyl-3-methylimidazolium ethyl sulfate	$Emim[EtSO_4]$	333	94.61	0.4570	Blanchard et al. 2001 [86]
1-Butyl,3-methyl-imidazolium hexafluorophosphate	$Bmim[PF_6]$	313	96.67	0.7290	Blanchard et al. 2001 [86]
1-Butyl-3-methylimidazolium acetate	$C_4mim[Ac]$	333.3	12.75	0.2510	Carvalho et al. 2009 [87]
		323.09	755.26	0.5990	Carvalho et al. 2009 [87]
1-Butyl-3-methylimidazolium trifluoroacetate	$C_4mim[TFA]$	293.43	9.79	0.2250	Carvalho et al. 2009 [87]
		293.59	436.25	0.6790	Carvalho et al. 2009 [87]

1-Butyl,3-methyl-imidazolium tetrafluoroborate	Bmim[BF$_4$]	303	10	0.1461	Galan-Sanchez 2008 [88]
		333	10	0.0895	Galan-Sanchez 2008 [88]
1-Octyl,3-methyl-imidazolium tetrafluoroborate	Omim[BF$_4$]	303	10	0.1873	Galan-Sanchez 2008 [88]
		333	10	0.1213	Galan-Sanchez 2008 [88]
1-Butyl,3-methyl-imidazolium dicyanamide	Bmim[DCA]	303	10	0.1434	Galan-Sanchez 2008 [88]
		333	10	0.0997	Galan-Sanchez 2008 [88]
1-Butyl-3-methylimidazolium thiocyanate	Bmim[SCN]	303	10	0.0978	Galan-Sanchez 2008 [88]
		333	10	0.0664	Galan-Sanchez 2008 [88]
1-Butyl,3-methyl-imidazolium hexafluorophosphate	Bmim[PF$_6$]	303	10	0.1662	Galan-Sanchez 2008 [88]
1-Butyl,3-methyl-imidazolium hexafluorophosphate	Bmim[PF$_6$]	333	10	0.1012	Galan-Sanchez 2008 [88]
1-Butyl-3-methylimidazolium methylsulfate	Bmim[MeSO$_4$]	303	10	0.1190	Galan-Sanchez 2008 [88]
		333	10	0.0733	Galan-Sanchez 2008 [88]

Compound	Abbreviation				Reference
1-N-Ethyl-3-mehylimidazolium bis(trifluoromethylsulfonyl)Imide	Emim[NTf$_2$]	303	10	0.2257	Galan-Sanchez 2008 [88]
		333	10	0.1446	Galan-Sanchez 2008 [88]
1-Butyl,3-methyl-imidazolium hexafluorophosphate	Bmim[PF$_6$]	298.15	6.66	0.122	Kim et al. 2005 [89]
1-Hexyl-3-methylimidazolium hexafluorophosphate	C$_6$mim[PF$_6$]	298.15	9.27	0.167	Kim et al. 2005 [89]
1-Ethyl-3-methylimidazolium tetrafluoroborate	Emim[BF$_4$]	298.15	8.75	0.106	Kim et al. 2005 [89]
1-Hexyl-3-methylimidazolium tetrafluoroborate	C$_6$mim[BF$_4$]	298.15	8.99	0.163	Kim et al. 2005 [89]
1-Ethyl-3-methylimidazolium bis(trifluoromethylsulfonyl)imide	Emim[Tf$_2$N]	298.15	9.03	0.209	Kim et al. 2005 [89]
1-Hexyl-3-methylimidazolium bis(trifluoromethylsulfonyl)imide	C$_6$mim[Tf$_2$N]	298.15	8.59	0.236	Kim et al. 2005 [89]
1-Ethyl-3-methylimidazolium trifluoromethane-sulfonate	C$_2$mim[TfO]	303.85	149	0.6260	Shin and Lee 2008 [90]
		303.85	15	0.2610	Shin and Lee 2008 [90]
1-Butyl-3-methylimidazolium trifluoromethane-sulfonate	C$_4$mim[TfO]	303.85	160	0.6720	Shin and Lee 2008 [90]
		303.85	11.5	0.2730	Shin and Lee 2008 [90]
1-Hexyl-3-methylimidazolium trifluoromethane-sulfonate	C$_6$mim[TfO]	303.85	80	0.7170	Shin and Lee 2008 [90]
		303.85	12.5	0.2880	Shin and Lee 2008 [90]

Compound	Abbreviation	Temperature		Solubility	Reference
1-Octyl-3-methylimidazolium trifluoromethane-sulfonate	$C_8mim[TfO]$	303.85	180	0.7410	Shin and Lee 2008 [90]
		303.85	15.8	0.3440	Shin and Lee 2008 [90]
1,3-Dimethylimidazolium methylphosphonate	Dmim[MP]	313.35	95	0.4750	Revelli et al. 2010 [91]
		313.45	34	0.1620	Revelli et al. 2010 [91]
1-Butyl,3-methyl-imidazolium tetrafluroborate	$Bmim[BF_4]$	293.65	73	0.6100	Revelli et al. 2010 [91]
		293.25	10.5	0.1410	Revelli et al. 2010 [91]
1-Butyl-3-methylimidazolium thiocyanate	Bmim[SCN]	313.65	99	0.4300	Revelli et al. 2010 [91]
		292.35	10.5	0.1260	Revelli et al. 2010 [91]
1-Ethyl-3-methylimidazolium trifluoroacetate	Emim[TFA]	298.1	19.99	0.2820	Yokozeki et al. 2008 [67]
1-Ethyl-3-methylimidazolium acetate	Emim[Ac]	298.1	19.99	0.4280	Yokozeki et al. 2008 [67]
1-Butyl-3-methylimidazolium trifluoroacetate	Bmim[TFA]	298.1	19.99	0.3010	Yokozeki et al. 2008 [67]
1-Butyl-3-methylimidazolium acetate	Bmim[Ac]	298.1	19.99	0.4550	Yokozeki et al. 2008 [67]
1-Ethyl-3-methyl-imidazolium bis(trifluoromethylsulfonyl)imide	$Emim[Tf_2N]$	298.1	19.99	0.3900	Yokozeki et al. 2008 [67]

1-Hexyl-3-methylimidazolium tris(pentafluoroethyl)trifluoro-phosphate	Hmim[FAP]	298.1	19.99	0.4930	Yokozeki et al. 2008 [67]
1-Hexyl-3-methylimidazolium bis(trifluoromethylsulfonyl)imide	Hmim[Tf$_2$N]	298.1	19.74	0.4330	Yokozeki et al. 2008 [67]
1-Butyl-3-methylimidazolium 1,1,2,2-tetrafluoroethanesulfonate	Bmim[TFES]	298	19.9	0.2850	Yokozeki et al. 2008 [67]
1-Butyl-3-methylimidazolium propionate	Bmim[PRO]	298.2	19.9	0.3900	Yokozeki et al. 2008 [67]
1-Butyl-3-methylimidazolium isobutyrate	Bmim[ISB]	298.2	20	0.4030	Yokozeki et al. 2008 [67]
1-Butyl-3-methylimidazolium trimethylacetate	Bmim[TMA]	298.1	19.9	0.4310	Yokozeki et al. 2008 [67]
1-Butyl-3-methylimidazolium levulinate	Bmim[LEV]	298.1	19.9	0.4600	Yokozeki et al. 2008 [67]
1-Butyl-3-methylimidazolium succinamate	Bmim[SUC]	298.1	19.9	0.2320	Yokozeki et al. 2008 [67]
Bis(1-butyl-3-methylimidazolium) iminodiacetate	Bmim$_{2)}$IDA]	298.1	19.9	0.3950	Yokozeki et al. 2008 [67]
1-Butyl-3-methylimidazolium iminoacetic acid acetate	Bmim[IAAc]	298.1	19.9	0.1910	Yokozeki et al. 2008 [67]
1-Hexyl-3-methylimidazolium tris(pentafluoroethyl)trifluorophosphate	Hmim[FEP]	283.5	17.99	0.5170	Zhang et al. 2008 [92]

Table 5: CO_2 solubility for ammonium ionic liquids

Ionic liquid	Acronym	T(K)	P(bar)	CO_2	References
Bis(2-hydroxyethyl)-ammonium acetate	(BHEAA)	298.15	15.15	0.1076	Kurnia et al. 2009 [78]
		298.15	5.48	0.0391	Kurnia et al. 2009 [78]
2-Hydroxy-N-(2-hydroxyethyl)-N-methylethanaminium acetate	(HHEMEA)	298.15	15.42	0.0761	Kurnia et al. 2009 [78]
		298.15	6.15	0.0300	Kurnia et al. 2009 [78]
Bis(2-hydroxyethyl)-ammonium lactate	(BHEAL)	298.15	15.12	0.0835	Kurnia et al. 2009 [78]
		298.15	3.46	0.0192	Kurnia et al. 2009 [78]
2-Hydroxy-N-(2-hydroxyethyl)-N-methylethanaminium lactate	(HHEMEL)	298.15	15.23	0.0776	Kurnia et al. 2009 [78]
		298.15	3.48	0.0179	Kurnia et al. 2009 [78]
2-Hydroxy ethyl ammonium formate	(HEF)	303	78.9	0.3083	Yuan et al. 2007 [93]
		303	4.4	0.0340	Yuan et al. 2007 [93]
2-Hydroxy ethyl ammonium acetate	(HEA)	303	90.1	0.4009	Yuan et al. 2007 [93]
		303	8.9	0.0687	Yuan et al. 2007 [93]
2-Hydroxy ethyl ammonium lactate	(HEL)	303	82	0.2422	Yuan et al. 2007 [93]
		303	7.8	0.0410	Yuan et al. 2007 [93]
Tri-(2-hydroxyethyl)-ammonium acetate	(THEAA)	303	82.5	0.2561	Yuan et al. 2007 [93]
		303	10.3	0.0534	Yuan et al. 2007 [93]
Tri-(2-hydroxyethyl)-ammonium lactate	(THEAL)	303	70.9	0.4617	Yuan et al. 2007 [93]
		303	9.6	0.1006	Yuan et al. 2007 [93]
2-(2-Hydroxyethoxy)-ammonium formate	(HEAF)	303	72.8	0.1907	Yuan et al. 2007 [93]
		303	6.6	0.0300	Yuan et al. 2007 [93]
2-(2-Hydroxyethoxy)-ammonium acetate	(HEAA)	303	65.7	0.4860	Yuan et al. 2007 [93]
		303	7.6	0.0889	Yuan et al. 2007 [93]
2-(2-Hydroxyethoxy)-ammonium lactate	(HEAL)	303	73.2	0.2640	Yuan et al. 2007 [93]
2-(2-Hydroxyethoxy)-ammonium lactate	(HEAL)	303	12.4	0.0704	Yuan et al. 2007 [93]

Due to certain limitations of conventional ionic liquid systems, where physical absorption takes place and high solubility is only seen at high pressures, numerous research groups have been developing the ILs designer character, by covalently tethering a functional group to either or both anion or cation. This resulting functionalized IL is capable of chemically binding to CO_2, adding chemical absorption to the capture mechanism.

[Bmim][Ac] has been found to be one of these RTILs in which a chemical complexion with CO_2 occurs [66]. In 2008, Yokozeki et al. [67] completed CO_2 solubility tests for 18 RTILs, eight of which showed chemical absorption mechanisms. They found that RTILs that show strong chemical absorption with CO_2 all contain the anion [X-COO]$^-$, that is, [Bmim][Ac], [Emim][Ac], [Bmim][PRO], [Bmim][IBS], [Bmim][TMA], and [Bmim][LEU]. Their results can be seen in Table 4. In general it is assumed that conventional RTILs with acidic or basic functionalities strongly influence the absorption of CO_2.

As discussed, RTILs sufficiently absorb CO_2 especially those containing CO_2-philic groups like fluorine. These are known as TSILs (task-specific ionic liquids). Widely researched TSILs are those with appended amine group, examples of which can be seen in Table 5. Bates and coworkers [68] synthesized the amine functionalized IL [pNH$_2$Bim] [Pf$_6$] and found it to chemically react with the CO_2. The CO_2 reacts with the amine on the IL, this then reacts with another amine and forms an ammonium carbamate double salt. This form of capture results in one CO_2 captured for every two ILs. This 1:2 capture mechanism is also observed on the molecular level with traditional aqueous amines. It is theoretically suggested that, when amines are tethered to the anion only, a 1:1 ratio can be met allowing a more efficient process.

Evidence has shown that TSILs have the ability to absorb CO_2 both chemically and physically. At low pressures (typically below 2 bars) chemical absorption takes place, in the same way as aqueous amines. After the majority of the chemical bonding have taken place, physical absorption dominates the capture mechanism; this is especially relevant at high pressures, whereas aqueous amines reach their absorption limits at low pressure. This shows how the absorption performances of TSILs with amine functionalities merge the characteristics of physical solvents with the attractive features of chemical solvents. In spite of TSILs showing greater CO_2 solubility than conventional RTILs, they tend to exhibit high viscosity in comparison to other commercially available absorbents. This poses a large problem for their implementation into large-scale platforms, as the heat required for absorption and regeneration would be a lot larger and energy intensive. In order to reduce the viscosity, some groups have combined mixtures of TSIL and RTIL. Bara et al. [69] dissolved their TSIL in a common RTIL, [C$_6$mim][Tf$_2$N]. Although the solution was stable and capable of absorbing in a

1:2 molar ratio, the viscosity was still high. As a whole TSILs and TSILs + RTILs are robust and have a high absorption capacity; however, they are limited by the intensive synthesis that is required, high viscosity, and the fact that the TSIL serves as both the capture material and the dispersant.

Instead of the direct incorporation of amino-functionalized anions and cations, some recent groups have reported using imidazolium-based RTILs with amines added in solution to act as the capture reagent. Camper and coworkers [70] first investigated this concept. They synthesized an [Rmim][Tf$_2$N] RTIL solution containing 16% v/v of MEA and found that this is capable of rapid and reversible capture of one mole of CO_2 per two moles of MEA at low CO_2 partial pressures. An MEA-carbamate was found to precipitate from the RTIL solution; this helps to drive the capture reaction. They have currently seen that this MEA-carbamate seems to be a consequence of the [Tf$_2$N] anion and does not occur in other [C$_n$mim] [X] RTILs.

Co_2 Solubility Results Reported by Various Experimental Groups

Tables 4, 5, 6, and 7 aim to provide a range of experimental data cited by various experimental groups, for peak CO_2 absorption values for different cation-based ILs. This can then be used to characterise an experimental system to ensure correct implementation and method.T (K) represents the system's temperature when measurements were recorded. P(bar) is the corresponding pressure of CO_2. XCO$_2$ is the solubility of CO_2 expressed as a mole fraction, that is, moles of CO_2 to moles of IL. The tables also attempt to show the effects of temperature and pressure on ILs as well as different cation and anion combinations.

Table 6: CO$_2$ solubility for phosphonium, pyridinium and pyrrolidinium ionic liquids

Ionic liquid	Acronym	T(K)	P(bar)	CO$_2$	References
N-Butylpyridinium tetrafluoroborate	N-BuPyBF$_4$	323	92.35	0.5810	Blanchard et al. 2001 [86]
Trihexyltetradecylphosphonium chloride	THTDPCl	302.55	149.95	0.8000	Carvalho et al. 2010 [94]
		313.27	5.17	0.2000	Carvalho et al. 2010 [94]
Trihexyltetradecylphosphonium bis(trifluoromethylsulfonyl)imide	THTDPNTf$_2$	296.58	721.85	0.8790	Carvalho et al. 2010 [94]
		293.2	6.12	0.3080	Carvalho et al. 2010 [94]
N-Butyl-4-methylpyridinium tetrafluoroborate	MeBuPyBF$_4$	303	10	0.1443	Galan-Sanchez 2008 [88]
		333	10	0.0961	Galan-Sanchez 2008 [88]
N-Butyl-3-Methylpyridinium dicyanamide	MeBuPyDCA	303	10	0.1436	Galan-Sanchez 2008 [88]
		333	10	0.0683	Galan-Sanchez 2008 [88]
N-Butyl-4-Methylpyridinium thiocyanate	MeBuPySCN	303	10	0.0962	Galan-Sanchez 2008 [88]
		333	10	0.0632	Galan-Sanchez 2008 [88]
1-Butyl-1-methylpyrrolidinium dicyanamide	MeBuPyrrDCA	303	10	0.1204	Galan-Sanchez 2008 [88]
		333	10	0.0613	Galan-Sanchez 2008 [88]

1-Butyl-1-Methylpyrrolidinium thiocyanate	MeBuPyrrSCN	303	10	0.0971	Galan-Sanchez 2008 [88]
		333	10	0.0608	Galan-Sanchez 2008 [88]
1-Butyl-1-methylpyrrolidinium trifluoroacetate	MeBuPyrrTFA	303	10	0.1674	Galan-Sanchez 2008 [88]
		333	10	0.1030	Galan-Sanchez 2008 [88]
Tetrabutylphosphonium formate	TBPFOR	298.1	19.9	0.3480	Yokozeki et al. 2008 [67]
1-Butyl-1-methylpyrrolidinium tri(pentafluoroethyl) trifluorophosphate	BmPyrr[FEP]	283.5	18.00	0.4980	Zhang et al. 2008 [92]

Table 7: CO$_2$ Solubility data for functionalized ionic liquids (TSILs)

Acronym	Functionalization	Anion	Cation	T(K)	P(bar)	xCO$_2$	References
APMimNTf$_2$	NH$_2$-cation	NTf$_2$	Im	303	10.00	0.27	Galan-Sanchez 2008 [88]
				343	10.00	0.18	
APMimDCA	NH$_2$-cation	DCA	Im	303	10.00	0.29	Galan-Sanchez 2008 [88]
APMimBF$_4$	NH$_2$-cation	BF$_4$	Im	303	10.00	0.32	Galan-Sanchez 2008 [88]
				343	10.00	0.36	
AEMPyrrBF$_4$	NH$_2$-cation	BF$_4$	Pyrr	303	10.00	0.28	Galan-Sanchez 2008 [88]
				333	10.00	0.24	
MeImNEt$_2$BF$_4$	NR$_3$-cation	BF$_4$	Im	303	4.00	0.09	Galan-Sanchez 2008 [88]
BmimTau	NH$_2$-Anion	Taureate	Im	333	10.00	0.43	Galan-Sanchez 2008 [88]
BmimGly	NH$_2$-Anion	Glycinate	Im	333	10.00	0.39	Galan-Sanchez 2008 [88]

EXPERIMENTAL AND MEASUREMENT TECHNIQUES

In order to integrate CO_2 separation techniques into large industrial systems, one needs to experimentally determine the ILs gas solubility in order to characterise the carrying capacity and selectiveness. These measurements can be accumulated via a number of experimental techniques, in which factors such as pressure and temperature can be controlled. The variety of techniques used for measuring solubility for high- and low-pressure phase equilibrium is vast and the naming of these techniques tends to vary from author to another. However all the techniques fall into two categories both of which are dependent on the equilibrium phases and mixture composition? If these two factors are unknown, measurements can be carried out analytically (analytical method); if the mixture is prepared with a precisely known composition, the synthetic method can be used. The experimental and measurement techniques reviewed here are gravimetric analysis, pressure drop method, and view-cell method and gas chromatography. All of which are being specifically applied to pure CO_2 solubility in ionic liquids. It is important to remember that impurities can occur in the gas and liquid, affecting the accuracy and precision of the results. Therefore degassing the liquid fully before analysis allows an accurate determination of the true solubility of the gas. This also relies on allowing true equilibrium conditions to be met between the gas and the liquid.

Gravimetric Analysis

Gravimetric analysis is an analytical method which describes the quantitative determination of, in this case, gas solubility by measuring the overall weight change of a sample during absorption. The gravimetric method is most commonly applied when the analyte is converted into a solid; however, as ILs are nonvolatile in nature and exhibit properties such as low vapour pressure, they can be used to a great effect with this method of analysis. Gravimetric gas analysers are used in laboratories conducting both fundamental studies into the physical properties of ILs and applications where the ability to measure gas solubility is of interest. The basic components of a typical gravimetric instrument can be seen in Figure 3.

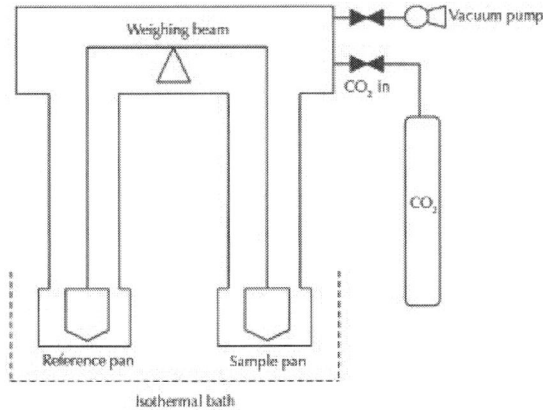

Figure 3: Basic components of a gravimetric system.

High-precision gravimetric instruments are commercially available, in the form of thermogravimetric microbalances. These allow in situ measurements of gas absorption that record the mass gain of a sample with a high-precision electro balance, which is capable of taking readings at high temperature and pressure. Also available are analysers which use magnetic suspension balances rather than an electrobalance [71]. The main difference between these two weighing systems is that, in magnetic suspension, the sample is weighed from the outside. Therefore the balance is not in physical contact with the high temperature and pressures subjected to the sample. This particular system is helpful when working with samples under extreme conditions. Petermann et al. [72] show the use and experimental setup of a magnetic suspension balance in conjunction with a volumetric determination method. The advantages of using magnetic suspension balances are also discussed by Dreisbach and Lösch [73].

Gravimetric analysis systems often measure gas solubility by recording isotherms, isobars, and kinetic sorption data, which can be output through a computer from which the system can be controlled. Hence when a sample is loaded, the operation of the instrument can be fully automated and programmed to carry out isothermal absorption and desorption measurements.

Due to gravimetric balances undergoing constant changes in temperature and pressure during measurements and the high sensitivity in which they operate, readings must be corrected for the changes in buoyant forces on the sample. In some apparatus, a counterweight side, which is symmetrical to the sample side, is used to minimise these effects. However they still need to be considered. Liu et al. [74] show a concise approach to calculate this.

A detailed experimental procedure using the gravimetric balance can be seen in [67]. Also measurements of CO_2 solubility for two imidazolium-based ILs using a thermogravimetric microbalance can be found in [75].

The Pressure Drop Method

The pressure drop method is a synthetic technique that is widely used in this scientific community and is also known as the isochoric method. In this instance the volume of the system is held constant, as well as the temperature, and the pressure difference is recorded during gas absorption into the sample. This method for working out gas solubility is practically suited for ILs as they have negligible vapour pressure, therefore ensuring that the gas phase remains pure, and therefore the assumption can be made that changes in pressure are due to gas sorption. From an initial measurement of pressure, temperature, and volume, and a final measurement of these variables at equilibrium, the amount of gas absorbed by the IL can be calculated. This calculation can be performed using an equation of state to convert all three variables into moles of gas.

The basic principles of this method are as follows: CO_2 gas is transferred into a reservoir of known volume and brought to a constant system temperature. An initial reading of pressure is measured. By using a PVT relation, the moles of CO_2 in the reservoir are calculated. The IL is loaded into an equilibrium cell/stainless steel reactor and equalized to system temperature. The CO_2 is then introduced to the ionic liquid and the pressure drop is recorded when the cell's pressure remains stable; this is the equilibrium point. From the pressure drop measured, the number of moles of CO_2 left in the gas phase can be calculated. The difference between CO_2 mole values corresponds to the amount of gas absorbed in the IL. A typical setup for the pressure drop method can be seen in Figure 4.

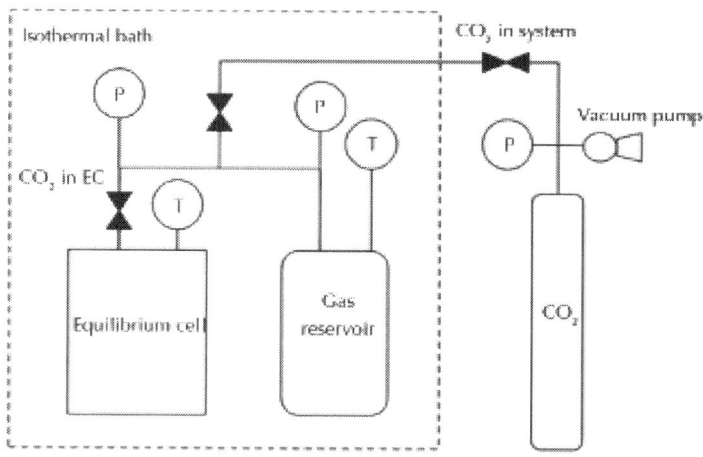

Figure 4: The pressure drop apparatus, where P and T correspond to pressure and temperature sensors.

The moles of dissolved CO_2 in the ionic liquid can be calculated by (1).

Number of CO_2 moles dissolved in the ionic liquid

$$n_{CO_2} = \frac{P_{initial} V_{GR}}{Z_{CO_2}(P_{initial}, T_{initial}) RT_{initial}} - \frac{P_{eq}(V_{tot} - V_{IL})}{Z_{CO_2}(P_{eq}, T_{eq}) RT_{eq}},$$

(1)

$P_{initial}$ and $T_{initial}$ are the initial pressure and temperature in the gas reservoir. P_{eq} and T_{eq} are the pressure and temperature at equilibrium in the equilibrium cell. V_{tot} is the total volume of the entire apparatus. V_{IL} is the volume of the ionic liquid, assumed to be constant. R is the ideal gas constant. Z_{CO} is the compressibility factor for CO_2; this modifies the ideal gas to account for real gas behaviour. A detailed experimental procedure and full calculations for CO_2 solubility measurements using the pressure drop method can be seen in [76, 77].

Further investigations that utilize this pressure drop method to derive gas solubility can be found where alternative experimental setups are shown [78–80].

View-Cell Methods

These involve the preparation of a mixture with a precisely known composition and then the observation of phase behavior inside an equilibrium cell, where measurements are recorded in the equilibrium state, that is, temperature and pressure. Synthetic methods consist of two main techniques, one being with a phase transition, and the other without. In synthetic methods with a phase transition a known amount of gas and IL is loaded into the equilibrium cell. The pressure is then varied at a constant temperature (or vice versa) until a second phase is formed, where the gas dissolves in the ionic liquid causing the vapor phase to diminish, whereby using different gas pressures, solubility can be worked out at various pressure, and temperatures.

In synthetic methods without a phase transition, equilibrium properties like temperature, pressure, density, cell volume, and gas/liquid phase volumes are measured, and the composition of the phase mixtures can be calculated in terms of moles or by a mass balance equation.

As can be seen in Figure 5, a pump releases CO_2 at a constant selected pressure and monitors the volume of CO_2 flowing into the system. The CO_2 is also heated to a constant temperature. By monitoring the volume, a known amount of CO_2 is then introduced to the high-pressure view cell, which contains a known amount of IL. In the case of non-phase transition, the amount of CO_2 absorbed is calculated by the difference in the amount of gas delivered to the cell and the amount of gas in the vapor phase. The amount of gas in the vapor phase can be calculated using a mass balance, shown in (2), coupled with an equation of state.

Equation to calculate amount of gas in the vapor phase:

$$m_g = m_{pump} - m_{lines} - m_{headspace} + m_{lines}^0 + m_{headspace}^0, \qquad (2)$$

Where m_g is the mass of CO_2 in the liquid phase, m_{pump} is the mass of CO_2 injected into the system, m_{lines} is the mass of CO_2 in the gas lines, connecting the pump to the equilibrium cell, is the mass of the gas in the headspace of the cell, m_{lines}^0 is the mass of the gas in the lines after venting the system, and is the mass of gas in the headspace, initially in the system after venting.

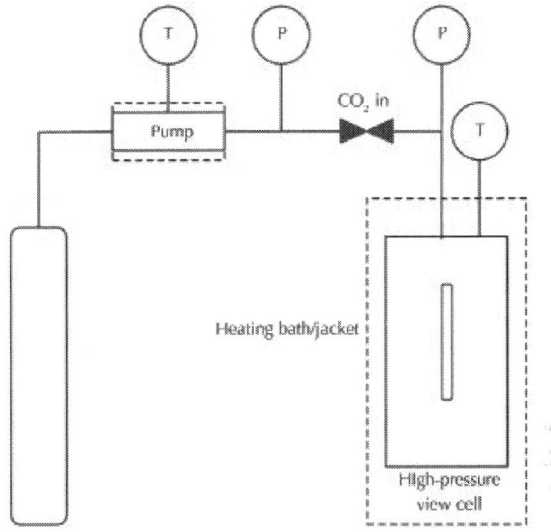

Figure 5: Scheme of a synthetic method setup.

A full experimental procedure using a synthetic method without phase transition and demonstrating the use of mass balancing to determine gas solubility is explained in the literature [30, 81, 82].

Further Techniques

Gas chromatography is an analytical method that boosts high precision and accuracy. When applied to measuring gas solubility in absorption media, the gas chromatograph is usually coupled with a high-pressure reactor cell in which a synthetic or pressure drop method is applied, and at equilibrium, a sample is taken and analyzed [83]. Solubility data from gas chromatography can be achieved by using an extractive technique; here the solvent (IL) is saturated with the solute (CO_2) and then coated on a column. Nitrogen, or any other nonabsorbing carrier gas, is directed on to the column in order to extract the CO_2 from the IL. The nitrogen is then analyzed in the gas chromatograph. This determines the amount of CO_2 removed (per amount of coating). A detailed method for applying gas chromatography can be found in the thesis by Wilbanks [84].

Other analytical techniques can be used in some cases to determine the solubility of specific gases; this may be in the form of a titration; this was demonstrated by Shen and Li [85] with aqueous amine solutions. However this has so far not been applied to ILs. Inline gas sensors also have the potential to be used. A possible scenario could involve linking an electrochemical sensor to measure the difference in CO_2 concentration of the vapor phase before and after equilibrium conditions.

Many advantages come from using gravimetric microbalances for solubility measurements. The ability to measure mass change to a high precision is helpful for a variety of reasons. When initially degassing the ionic liquid sample, being able to measure mass decrease allows the experimentalist to see when a constant mass value has been reached, thus allowing the assumption that full degassing has occurred. Also, the mass reading is very important to ensure equilibrium conditions once the CO_2 has been introduced. Equilibrium is reached when the mass change is zero. In situ gravimetric balances, that is, when the balance is enclosed in the measuring gas, are limited to lower pressures and temperatures. Disadvantages of gravimetric systems are mainly due to their high retail price, making them impractical for small projects where the funding is restricted.

In comparison to the gravimetric analysis, the pressure drop and synthetic methods are much simpler in design. As samples of any size can be investigated with these methods, a high sensitivity can generally be achieved, however, not a high accuracy. The most significant errors in the pressure drop and synthetic methods are the error calculations of dead space; for gravimetric methods, it is the determination of buoyancy forces. In pressure drop and synthetic methods, the two variables, pressure and gas absorbed, are determined by the pressure sensors and calibrated volumes; this can result in measuring error which is added on each step of the absorption isotherm. With the gravimetric method, all of the variables, temperature, gas pressure, and absorbed gas, are measured independently and the absorption pressure is monitored at each step of the isotherm.

CONCLUSIONS

The versatility and inherent advantages of ionic liquids in the process of CO_2 capture are giving rise to a promising and expansive field. Their potential as physical absorbents is highly attractive, although at present their capture rate is not to the same scale as current aqueous amine technologies; the fact that amines for CO_2 capture have been developed through many years and that ILs are a new research field leaves room for further research and improvement.

Solubility data of CO_2 in different imidazolium-based ionic liquids are the most often found in the literature. This is especially the case for bmim[BF$_4$] and bmim[PF$_6$], because these ionic liquids were among the first ones commercially available. Therefore an abundant amount of previous data is available and allows the validation of subsequent experimental procedures. Although commercially available, the price of these ionic liquids remains high. Quaternary ammonium and tetra alkylphosphonium bases provide a cheaper alternative. In comparison the synthesis process of these ionic liquids is simpler and the raw materials are accessible. However the lack of experimental data with these solvents means that they are constantly overshadowed.

Although experimental data on CO_2 solubility in ionic liquids is available in the literature, more is needed for process design. Here several different methods have been presented in order to obtain this data. These include gravimetric analysis, pressure drop, and synthetic methods, all of which are particularly well suited for the measurement of gases in nonvolatile liquids. In terms of solubility data measurements, gravimetric balances offer the simplest and most precise route; however, their general high prices make them impractical for small research groups conducting initial experiments with ionic liquids. Pressure drop and synthetic methods provide a cheaper alternative and do not need sampling. However these methods depend on the models used to calculate the thermodynamic properties and phase equilibrium. It is important to observe that for some thermodynamic properties, such as excess molar enthalpy, research groups use a test system to check their equipment and methods accuracy. In the case of gas-liquid solubility, however, there is no test system, especially at elevated temperatures and pressures.

The main challenges affecting ionic liquids as green solvent for CO_2 capture are availability, cost, purity, and compatibility. These challenges are faced at present on a laboratory scale and must have solutions before expanding to industry. At present the advantages and disadvantages of ionic liquids and amines seem to be equally balanced. The main criteria for ideal CO_2 capture mechanisms are high CO_2 solubility, low energy input for regeneration, low cost, long-term reusability, and being environmentally benign. At the moment amines have the advantage of having high CO_2 solubility and being of low cost. However due to the vast number of ionic liquids that can be developed and different ways in which they can be synthesized, the potential is there. Moreover through increasing research and commercialization of ionic liquids in other areas of industry, the cost is set to decrease.

ACKNOWLEDGMENTS

This work has been supported by FEDER, ACCIO, and the Government of Catalonia (Funding TECRD12-1-0010).

REFERENCES

1. B. Metz, O. Davidson, H. deConinck, M. Loos, and L. Meyer, IPCC special report on carbon dioxide capture and storage, prepared by working group III of the intergovernmental panel on climate change, Cambridge University Press, New York, NY, USA, 2005.

2. Carbon Capture and Storage in Industrial Applications: Technology Synthesis Report Working Paper—November 2010, United Nations Industrial Development Organization.

3. A. Stangeland, "A model for the CO_2 capture potential," International Journal of Greenhouse Gas Control, vol. 1, no. 4, pp. 418–429, 2007.

4. S. T. Brennan, R. C. Burruss, M. D. Merrill, P. A. Freeman, and L. F. Ruppert, "A probabilistic assessment methodology for the evaluation of geologic carbon dioxide storage," U.S. Geological Survey Open-File Report 2010-1127, 2010.

5.   R. P. Hepple and S. M. Benson, "Geologic storage of carbon dioxide as a climate change mitigation strategy: performance requirements and the implications of surface seepage," Environmental Geology, vol. 47, no. 4, pp. 576–585, 2005.

6.   S. Holloway, "An overview of the Joule II project: the underground disposal of carbon dioxide," Energy Conversion and Management, vol. 37, no. 6–8, pp. 1149–1154, 1996.

7.   E. T. Sundquist, R. C. Burruss, S. P. Faulkner et al., "Carbon sequestration to mitigate climate change," U.S. Geological Survey, Fact Sheet 2008-3097, 2008.

8.   M. Finkenrath, Cost and Performance of Carbon Dioxide Capture from Power Generation, International Energy Agency, 2011.

9.   R. S. Haszeldine, "Carbon capture and storage: how green can black be?" Science, vol. 325, no. 5948, pp. 1647–1652, 2009.

10. T. Kuramochi, A. Faaij, A. Ramírez, and W. Turkenburg, "Prospects for cost-effective post-combustion CO_2 capture from industrial CHPs," International Journal of Greenhouse Gas Control, vol. 4, no. 3, pp. 511–524, 2010.

11. H. Manchao, L. Sousa, R. Sousa, A. Gomes, E. Vargas Jr., and Z. Na, "Risk assessment of CO_2 injection processes and storage in carboniferous formations: a review," Journal of Rock Mechanics and Geotechnical Engineering, vol. 3, no. 1, pp. 39–56, 2011.

12. S.-E. Park, J.-S. Chang, and K.-W. Lee, Eds., Carbon Dioxide Utilization For Global Sustainability, Proceedings of 7th the International Conference on Carbon Dioxide Utilization, vol. 153, 2004.

13. I. Omae, "Aspects of carbon dioxide utilization," Catalysis Today, vol. 115, no. 1–4, pp. 33–52, 2006.

14. D. M. D'Alessandro, B. Smit, and J. R. Long, "Carbon dioxide capture: prospects for new materials," Angewandte Chemie, vol. 49, no. 35, pp. 6058–6082, 2010.

15. X. Gui, Z. Tang, and W. Fei, "CO_2 capture with physical solvent dimethyl carbonate at high pressures," Journal of Chemical and Engineering Data, vol. 55, no. 9, pp. 3736–3741, 2010.

16. A. L. Chaffee, G. P. Knowles, Z. Liang, J. Zhang, P. Xiao, and P. A. Webley, "CO_2 capture by adsorption: materials and process

development," International Journal of Greenhouse Gas Control, vol. 1, no. 1, pp. 11–18, 2007.

17. C. A. Scholes, S. E. Kentish, and G. W. Stevens, "Carbon dioxide separation through polymeric membrane systems for flue gas applications," Recent Patents on Chemical Engineering, vol. 1, no. 1, pp. 52–66, 2008.

18. S. Burt, A. Baxte, and L. Baxter, Cryogenic CO_2 Capture to Control Climate Change Emissions, Brigham Young University, Provo, Utah, USA, 84602 Sustainable Energy Solutions Orem, 84058, 2010.

19. L. Faramarzi, G. M. Kontogeorgis, M. L. Michelsen, K. Thomsen, and E. H. Stenby, "Absorber model for CO_2 capture by monoethanolamine," Industrial and Engineering Chemistry Research, vol. 49, no. 8, pp. 3751–3759, 2010.

20. P. J. G. Huttenhuis, N. J. Agrawal, E. Solbraa, and G. F. Versteeg, "The solubility of carbon dioxide in aqueous N-methyldiethanolamine solutions," Fluid Phase Equilibria, vol. 264, no. 1-2, pp. 99–112, 2008.

21. M. L. Kennard and A. Melsen, "Solubility of carbon dioxide in aqueous diethanolamine solutions at elevated temperatures and pressures," Journal of Chemical and Engineering Data, vol. 29, no. 3, pp. 309–312, 1984.

22. H. Lepaumier, S. Martin, D. Picq, B. Delfort, and P. Carrette, "New amines for CO_2 capture. III. Effect of alkyl chain length between amine functions on polyamines degradation," Industrial and Engineering Chemistry Research, vol. 49, no. 10, pp. 4553–4560, 2010.

23. G. T. Rochelle, "Amine scrubbing for CO_2 capture," Science, vol. 325, no. 5948, pp. 1652–1654, 2009.

24. G. Puxty, R. Rowland, A. Allport et al., "Carbon dioxide postcombustion capture: a novel screening study of the carbon dioxide absorption performance of 76 amines," Environmental Science and Technology, vol. 43, no. 16, pp. 6427–6433, 2009.

25. F. Closmann, T. Ngugen, and G. T. Rochelle, "MDEA/Piperazine as a solvent for CO_2 capture," Energy Procedia, vol. 1, no. 1, pp. 1351–1357, 2009.

26. P. W. F. Riemer and W. G. Ormerod, "International perspectives and the results of carbon dioxide capture disposal and utilisation studies," Energy Conversion and Management, vol. 36, no. 6–9, pp. 813–818, 1995.

27. M. Karl, R. F. Wright, T. F. Berglen, and B. Denby, "Worst case scenario study to assess the environmental impact of amine emissions from a CO_2 capture plant," International Journal of Greenhouse Gas Control, vol. 5, no. 3, pp. 439–447, 2011.

28. J. Kittel, R. Idem, D. Gelowitz, P. Tontiwachwuthikul, G. Parrain, and A. Bonneau, "Corrosion in MEA units for CO_2 capture: pilot plant studies," Energy Procedia, vol. 1, no. 1, pp. 791–797, 2009.

29. A. B. Rao and E. S. Rubin, "A technical, economic, and environmental assessment of amine-based CO_2 capture technology for power plant greenhouse gas control," Environmental Science and Technology, vol. 36, no. 20, pp. 4467–4475, 2002.

30. G. Mahrer and D. Tuma, "Gas solubility (and related high-pressure phenomena) in systems with ionic liquids," in Ionic Liquids: From Knowledge to Application, N. V. Plechkova, R. D. Rogers, and K. R. Seddon, Eds., chapter 1, pp. 1–20.

31. J. D. Holbrey and K. R. Seddon, "Ionic liquids," Clean Products and Processes, vol. 1, no. 4, pp. 223–236, 1999.

32. K. R. Seddon, "Ionic liquids: a taste of the future," Nature Materials, vol. 2, no. 6, pp. 363–365, 2003.

33. M. Galiński, A. Lewandowski, and I. Stepniak, "Ionic liquids as electrolytes," Electrochimica Acta, vol. 51, no. 26, pp. 5567–5580, 2006.

34. P. Wang, S. M. Zakeeruddin, J.-E. Moser, and M. Grätzel, "A new ionic liquid electrolyte enhances the conversion efficiency of dye-sensitized solar cells," The Journal of Physical Chemistry B, vol. 107, no. 48, pp. 13280–13285, 2003.

35. A. E. Jimenez, M. D. Bermudez, F. J. Carrion, and G. Martınez-Nicolas, "Room temperature ionic liquids as lubricant additives in steel-aluminium contacts: influence of sliding velocity, normal load and temperature," Wear, vol. 261, no. 3-4, pp. 347–359, 2006.

36. A. P. Abbott, K. J. McKenzie, and K. S. Ryder, "Electropolishing and electroplating of metals using ionic liquids based on choline

chloride," in Ionic Liquids IV, vol. 975 of ACS Symposium Series, chapter 13, pp. 186–197, 2007.

37. S. S. Y. Tan and D. R. Macfarlane, "Ionic liquids in biomass processing," Topics in Current Chemistry, vol. 290, pp. 311–339, 2009.

38. S. V. Dzyuba, K. D. Kollar, and S. S. Sabnis, "Synthesis of imidazolium room-temperature ionic liquids exploring green chemistry and click chemistry paradigms in undergraduate organic chemistry laboratory," Journal of Chemical Education, vol. 86, no. 7, pp. 856–858, 2009.

39. A. Stark, D. Ott, D. Kralisch, G. Kreisel, and B. Ondruschka, "Ionic liquids and green chemistry: a lab experiment," Journal of Chemical Education, vol. 87, no. 2, pp. 196–201, 2010.

40. T. Welton, "Room-temperature ionic liquids. Solvents for synthesis and catalysis," Chemical Reviews, vol. 99, no. 8, pp. 2071–2083, 1999.

41. M. J. Earle and K. R. Seddon, "Ionic liquids. Green solvents for the future," Pure and Applied Chemistry, vol. 72, no. 7, pp. 1391–1398, 2000.

42. C. M. Gordon, "New developments in catalysis using ionic liquids," Applied Catalysis A, vol. 222, no. 1-2, pp. 101–117, 2001.

43. J. D. Holbrey, "Industrial applications of ionic liquids," Chimica Oggi, vol. 22, no. 6, pp. 35–37, 2004.

44. Q. Gan, D. Rooney, and Y. Zou, "Supported ionic liquid membranes in nanopore structure for gas separation and transport studies," Desalination, vol. 199, no. 1–3, pp. 535–537, 2006.

45. C. P. Fredlake, J. M. Crosthwaite, D. G. Hert, S. N. V. K. Aki, and J. F. Brennecke, "Thermophysical properties of imidazolium-based ionic liquids," Journal of Chemical and Engineering Data, vol. 49, no. 4, pp. 954–964, 2004.

46. K. R. Seddon, A. Stark, and M. J. Torres, "Viscosity and density of 1-alkyl-3-methylimidazolium ionic liquids," in Clean Solvents, M. Abraham and L. Moens, Eds., pp. 34–49, American Chemical Society, Washington, DC, USA, 2002.

47. R. A. Mantz and P. C. Trulove, "Viscosity and density of ionic liquids. Physicochemical properties," in Ionic Liquids in Synthesis,

P. Wasserscheid and T. Welton, Eds., pp. 56–68, Wiley-VCH Verlag GmbH & Co. KGaA, Weinheim, Germany, 2nd edition, 2008.

48. J. W. Magee, "Heat capacity and enthalpy of fusion for 1-butyl-3-methyl-imidazolium hexafluorophosphate," in Proceedings of the 17th IUPAC Conference on Chemical Thermodynamics (ICCT ‹02), Rostock, Germany, 2002.

49. M. Kosmulski, J. Gustafsson, and J. B. Rosenholm, "Thermal stability of low temperature ionic liquids revisited," Thermochimica Acta, vol. 412, no. 1-2, pp. 47–53, 2004.

50. G. Law and P. R. Watson, "Surface tension measurements of N-alkylimidazolium ionic liquids,"Langmuir, vol. 17, no. 20, pp. 6138–6141, 2001.

51. A. S. Wells and V. T. Coombe, "On the freshwater ecotoxicity and biodegradation properties of some common ionic liquids," Organic Process Research and Development, vol. 10, no. 4, pp. 794–798, 2006.

52. T. D. Landry, K. Brooks, D. Poche, and M. Woolhiser, "Acute toxicity profile of 1-butyl-3-methylimidazolium chloride," Bulletin of Environmental Contamination and Toxicology, vol. 74, no. 3, pp. 559–565, 2005.

53. I. Perissi, U. Bardi, S. Caporali, and A. Lavacchi, "High temperature corrosion properties of ionic liquids," Corrosion Science, vol. 48, no. 9, pp. 2349–2362, 2006.

54. M. Uerdingen, C. Treber, M. Balser, G. Schmitt, and C. Werner, "Corrosion behaviour of ionic liquids,"Green Chemistry, vol. 7, no. 5, pp. 321–325, 2005.

55. P. T. Anastas and J. C. Warner, Green Chemistry: Theory and Practice, Oxford University Press, New York, NY, USA, 1998.

56. P. Scovazzo, J. Kieft, D. A. Finan, C. Koval, D. DuBois, and R. Noble, "Gas separations using non-hexafluorophosphate [PF$_6$]$^-$ anion supported ionic liquid membranes," Journal of Membrane Science, vol. 238, no. 1-2, pp. 57–63, 2004.

57. R. E. Baltus, B. H. Culbertson, S. Dai, H. Luo, and D. W. DePaoli, "Low-pressure solubility of carbon dioxide in room-temperature ionic liquids measured with a quartz crystal microbalance," The Journal of Physical Chemistry B, vol. 108, no. 2, pp. 721–727, 2004.

58. Á. P.-S. Kamps, D. Tuma, J. Xia, and G. Maurer, "Solubility of CO_2 in the ionic liquid [bmim][PF$_6$]," Journal of Chemical and Engineering Data, vol. 48, no. 3, pp. 746–749, 2003.

59. B.-C. Lee and S. L. Outcalt, "Solubilities of gases in the ionic liquid 1-n-butyl-3-methylimidazolium   bis(trifluoromethylsulfonyl) imide," Journal of Chemical and Engineering Data, vol. 51, no. 3, pp. 892–897, 2006.

60. M. J. Muldoon, S. N. V. K. Aki, J. L. Anderson, J. K. Dixon, and J. F. Brennecke, "Improving carbon dioxide solubility in ionic liquids," The Journal of Physical Chemistry B, vol. 111, no. 30, pp. 9001–9009, 2007.

61. C. Cadena, J. L. Anthony, J. K. Shah, T. I. Morrow, J. F. Brennecke, and E. J. Maginn, "Why is CO_2 so soluble in imidazolium-based ionic liquids?" Journal of the American Chemical Society, vol. 126, no. 16, pp. 5300–5308, 2004.

62. S. N. V. K. Aki, B. R. Mellein, E. M. Saurer, and J. F. Brennecke, "High-pressure phase behavior of carbon dioxide with imidazolium-based ionic liquids," The Journal of Physical Chemistry B, vol. 108, no. 52, pp. 20355–20365, 2004.

63. P. J. Carvalho, V. H. Álvarez, I. M. Marrucho, M. Aznar, and J. A. P. Coutinho, "High carbon dioxide solubilities in trihexyltetradecylphosphonium-based ionic liquids," Journal of Supercritical Fluids, vol. 52, no. 3, pp. 258–265, 2010.

64. S. G. Kazarian, B. J. Briscoe, and T. Welton, "Combining ionic liquids and supercritical fluids: in situATR-IR study of CO_2 dissolved in two ionic liquids at high pressures," Chemical Communications, no. 20, pp. 2047–2048, 2000.

65. E. J. Beckman, "A challenge for green chemistry: designing molecules that readily dissolve in carbon dioxide," Chemical Communications, vol. 10, no. 17, pp. 1885–1888, 2004.

66. M. B. Shiflett and A. Yokozeki, "Solubilities and diffusivities of carbon dioxide in ionic liquids: [bmim][PF$_6$] and [bmim][BF$_4$]," Industrial and Engineering Chemistry Research, vol. 44, no. 12, pp. 4453–4464, 2005.

67. A. Yokozeki, M. B. Shiflett, C. P. Junk, L. M. Grieco, and T. Foo, "Physical and chemical absorptions of carbon dioxide in room-temperature ionic liquids," The Journal of Physical Chemistry B, vol. 112, no. 51, pp. 16654–16663, 2008.

68. E. D. Bates, R. D. Mayton, I. H. Ntai, et al., "CO$_2$ 2 capture by a task-specific ionic liquid," Journal of the American Chemical Society, vol. 124, no. 6, pp. 926–927, 2002.

69. J. E. Bara, D. E. Camper, D. L. Gin, and R. D. Noble, "Room-temperature ionic liquids and composite materials: platform technologies for CO$_2$ capture," Accounts of Chemical Research, vol. 43, no. 1, pp. 152–159, 2010.

70. D. Camper, J. E. Bara, D. L. Gin, and R. Noble, "Room-temperature ionic liquid-amine solutions: tunable solvents for efficient and reversible capture of CO$_2$," Industrial & Engineering Chemistry Research, vol. 47, no. 21, pp. 8496–8498, 2008.

71. R. Kleinrahm and W. Wagner, "Measurement and correlation of the equilibrium liquid and vapour densities and the vapour pressure along the coexistence curve of methane," The Journal of Chemical Thermodynamics, vol. 18, no. 8, pp. 739–760, 1986.

72. M. Petermann, T. Weissert, S. Kareth, H. W. Lösch, and F. Dreisbach, "New instrument to measure the selective sorption of gas mixtures under high pressures," Journal of Supercritical Fluids, vol. 45, no. 2, pp. 156–160, 2008.

73. F. Dreisbach and H. W. Lösch, "Magnetic suspension balance for simultaneous measurement of a sample and the density of the measuring fluid," Journal of Thermal Analysis and Calorimetry, vol. 62, no. 2, pp. 515–521, 2000.

74. H. Liu, J. Huang, and P. Pendleton, "Experimental and modelling study of CO$_2$ absorption in ionic liquids containing Zn (II) ions," Energy Procedia, vol. 4, pp. 59–66, 2011.

75. A. N. Soriano, B. T. Doma Jr., and M.-H. Li, "Solubility of carbon dioxide in 1-ethyl-3-methylimidazolium tetrafluoroborate," Journal of Chemical and Engineering Data, vol. 53, no. 11, pp. 2550–2555, 2008.

76. J. Palgunadi, J. E. Kang, D. Q. Nguyen et al., "Solubility of CO$_2$ in dialkylimidazolium dialkylphosphate ionic liquids," Thermochimica Acta, vol. 494, no. 1-2, pp. 94–98, 2009.

77. J. Palgunadi, J. E. Kang, M. Cheong, H. Kim, H. Lee, and H. S. Kim, "Fluorine-free imidazolium-based ionic liquids with a phosphorous-containing anion as potential CO$_2$ absorbents," Bulletin of the Korean Chemical Society, vol. 30, no. 8, pp. 1749–1754, 2009.

78. K. A. Kurnia, F. Harris, C. D. Wilfred, M. I. Abdul Mutalib, and T. Murugesan, "Thermodynamic properties of CO_2 absorption in hydroxyl ammonium ionic liquids at pressures of (100–1600) kPa," The Journal of Chemical Thermodynamics, vol. 41, no. 10, pp. 1069–1073, 2009.

79. D. Camper, P. Scovazzo, C. Koval, and R. Noble, "Gas solubilities in room-temperature ionic liquids,"Industrial and Engineering Chemistry Research, vol. 43, no. 12, pp. 3049–3054, 2004.

80. F. C. Gomes, "Solubility of carbon dioxide, ethane, methane, oxygen, nitrogen, hydrogen, argon, and carbon monoxide in 1-butyl-3-methylimidazolium tetrafluoroborate between temperatures 283 K and 343 K and at pressures close to atmospheric," The Journal of Chemical Thermodynamics, vol. 38, no. 4, pp. 490–502, 2006.

81. W. Ren and A. M. Scurto, "High-pressure phase equilibria with compressed gases," Review of Scientific Instruments, vol. 78, no. 12, Article ID 125104, 7 pages, 2007.

82. A.-L. Revelli, F. Mutelet, and J.-N. Jaubert, "High carbon dioxide solubilities in imidazolium-based ionic liquids and in poly(ethylene glycol) dimethyl ether," The Journal of Physical Chemistry B, vol. 114, no. 40, pp. 12908–12913, 2010.

83. Z. Lei, J. Yuan, and J. Zhu, "Solubility of CO_2 in propanone, 1-ethyl-3-methylimidazolium tetrafluoroborate, and their mixtures," Journal of Chemical and Engineering Data, vol. 55, no. 10, pp. 4190–4194, 2010.

84. K. E. Wilbanks, Phase behavior of carbon dioxide and oxygen in the ionic liquid 1-hexyl-3-methylimidazolium bis(trifluoromethylsulfonyl) imide [M.S. thesis], 2007.

85. K. P. Shen and M. H. Li, "Solubility of carbon dioxide in aqueous mixtures of monoethanolamine with methyldiethanolamine," Journal of Chemical and Engineering Data, vol. 37, no. 1, pp. 96–100, 1992.

86. L. A. Blanchard, Z. Gu, and J. F. Brennecke, "High-pressure phase behavior of ionic liquid/CO_2systems," The Journal of Physical Chemistry B, vol. 105, no. 12, pp. 2437–2444, 2001.

87. P. J. Carvalho, V. H. Álvarez, B. Schröder et al., "Specific solvation interactions of CO_2 on acetate and trifluoroacetate imidazolium

based ionic liquids at high pressures," The Journal of Physical Chemistry B, vol. 113, no. 19, pp. 6803–6812, 2009.

88. L. M. Galan-Sanchez, Functionalised ionic liquids, absorption solvents for CO_2 and olefin separation [Ph.D. thesis], 2008.

89. Y. S. Kim, W. Y. Choi, J. H. Jang, K.-P. Yoo, and C. S. Lee, "Solubility measurement and prediction of carbon dioxide in ionic liquids," Fluid Phase Equilibria, vol. 228-229, pp. 439–445, 2005.

90. E.-K. Shin and B.-C. Lee, "High-pressure phase behavior of carbon dioxide with ionic liquids: 1-alkyl-3-methylimidazolium trifluoromethanesulfonate," Journal of Chemical and Engineering Data, vol. 53, no. 12, pp. 2728–2734, 2008.

91. A.-L. Revelli, F. Mutelet, and J.-N. Jaubert, "High carbon dioxide solubilities in imidazolium-based ionic liquids and in poly(ethylene glycol) dimethyl ether," The Journal of Physical Chemistry B, vol. 114, no. 40, pp. 12908–12913, 2010.

92. X. Zhang, Z. Liu, and W. Wang, "Screening of ionic liquids to capture CO_2 by COSMO-RS and experiments," AIChE Journal, vol. 54, no. 10, pp. 2717–2728, 2008.

93. X. Yuan, S. Zhang, J. Liu, and X. Lu, "Solubilities of CO_2 in hydroxyl ammonium ionic liquids at elevated pressures," Fluid Phase Equilibria, vol. 257, no. 2, pp. 195–200, 2007.

94. P. J. Carvalho, V. H. Alvarez, I. M. Marrucho, M. Aznar, and J. A. P. Coutinho, "High carbon dioxide solubilities in trihexyltetradecylphosphonium-based ionic liquids," The Journal of Supercritical Fluids, vol. 252, no. 3, pp. 258–265, 2010.

Chapter 8

Compressive Strength Properties of Natural Gas Hydrate Pellet by Continuous Extrusion from a Twin-Roll System

Yun-Hoo Lee, Bong-Hwan Koh, Heung Soo Kim, and Myung Ho Song

Department of Mechanical, Robotics and Energy Engineering, Dongguk University-Seoul, 30 Pildong-ro, 1-gil, Jung-gu, Seoul 100-715, Republic of Korea

ABSTRACT

This study investigates the compressive strength of natural gas hydrate (NGH) pellet strip extruded from die holes of a twin-roll press for continuous pelletizing (TPCP). The lab-scale TPCP was newly developed, where NGH powder was continuously fed and extruded into strip-type pellet between twin rolls. The system was specifically designed for future expansion towards mass production of solid form NGH. It is shown that the compressive strength of NGH pellet strip heavily depends on parameters in the extrusion process, such as feeding pressure, pressure ratio, and rotational speed. The mechanism of TPCP, along with the compressive strength and density of pellets, is discussed in terms of its feasibility for producing NGH pellets in the future.

INTRODUCTION

Natural gas hydrate (NGH), widely known for its self-preservation effect [1], remains in a metastable state or quasiequilibrium under atmospheric pressure and subzero temperature (below −20°C). This property provides a solid background for the massive-scale exploitation of NGH and the development of pertinent technologies in storage, handling, and transportation. In this context, some of the economic benefits from the exploitation of NGH over conventional liquefied natural gas (LNG) have aroused significant global attention in recent years. Because NGH conveniently captures gases up to 170 times the amount of its own volume, the storage capacity per unit volume of pellet-type hydrate turns out to be much higher than that of compressed natural gas (CNG) [2–4]. Therefore, transporting natural gas in the form of NGH pellets has a robust benefit over LNG, in terms of price and reliability. The technology related to gas-to-solid (GTS) transition, for transporting natural gas produced from stranded small gas wells in the form of hydrate, becomes a promising alternative for conventional LNG transportation systems [5, 6]. However, if the methane concentration is not rich enough, the self-preservation effect of NGH pellet may disappear. Thus, methane must be firmly trapped inside of the NGH pellet as a surrounding gas, if the pellet needs to be transported for prolonged time [7]. In order to fully exploit the advantage of the self-preservation effect, the technology related to GTS and pelletizing NGH becomes a critical research issue for many of the natural gas industries. Considering the massive scale of gas transportation, the extrusion-type method is the most effective for the continuous production of NGH pellets. Accordingly, to tackle the issue of NGH mass production, laboratory-scale continuous pelletizer for pilot-plants has been developed and discussed by numerous researchers in recent years [8].

In a previous study, Choi and Koh reported that the strength of ice pellet was dependent on the compressive pressure [9]. They developed capsule and cylinder shape vertical pelletizer, to investigate the effect of compressive pressure on the strength of ice pellet. Ko et al. and Shim also showed that the rotational speed and starting angle of pelletizer affect the physical properties of pellets [10, 11]. Also, Jung et al. recently investigated a hopper system of ball-type NGH pellets using the response surface method, to find an optimal design condition for hopper discharge [12]. Therefore, the strength of NGH pellet depends on the design parameters of the pelletizer, such as the size and surface friction of the disk, feeder mechanism, temperature conditions, and rotational speed. The study of these design parameters is a key issue for the development of the mass production of NGH pellet and the solid transport system of natural gas.

In this paper, we developed an extrusion-type twin-roll Press for continuous pelletizing system or TPCP, which is specifically designed for continuous production of NGH pellet. The TPCP is a laboratory-scale NGH pelletizing device. It is a gear-type system with two adjacent rolls, which push together and extrude hydrate powder through the outlet hole. The rotating twin-roll and overhead feeding weight create enough squeezing pressure to form hydrate pellet. The feeding hydrate powder was compounded through the lab-scale NGH reactor in advance. Particles of powder having diameter less than 1 mm were continuously fed into the pelletizing system. By using TPCP, we successfully extruded a long and thin rod-shaped NGH pellet. Also, we are able to measure the compressive strength of natural gas hydrate pellet according to the basic design parameters of pelletizer, such as feeder pressure, rotational speed of the disk, and starting angle of the pelletizer. Experimental results show that the mechanical strength of NGH pellet extruded by the TPCP system mainly depends on the rotational speed of the twin roll and overhead feeder pressure. This investigation result can be used as a background data for designing a mass production system of NGH pellet and loading/unloading mechanism for NGH ships in the near future.

EXPERIMENT

Hydrate Formation

In order to fabricate NGH pellets through the TPCP system, the feeding material or NGH powder compound needs to be formed through hydrate reactor. Figure 1 shows the schematics and process diagram of a laboratory-scale hydrate forming reactor. In order to meet the hydrate forming condition, a constant temperature bath was controlled to sustain a temperature level of −0.5°C. Also, 15% of ethylene glycol was added to the cooling water to prevent freezing, and coolant was circulated around the outer surface of the reactor. Before forming the hydrate powder compound, methane gas was filled up to 120 bar at the accumulator along with 2 kg water, and dozens of heavy stainless balls were added in the reactor. To remove the air in the reactor, methane gas was pressured up to 3 bar in the reactor, and the pressure was reduced down to atmospheric pressure through the vent line. This process was repeated three times, to completely remove air in the reactor. After air was removed, the reactor was pressured with methane gas up to 60 bar in the accumulator. Then, a motor-driven impeller was rotated to stir the methane gas, water, and stainless balls. During the process, constant-pressured methane gas was supplied to the reactor through a pneumatic regulator operated by a flow management program, and the pressure in the reactor was kept up to 60 bar. After 80 minutes, the hydrate powder was being formed in the reactor, and we could extract the powder at the constant temperature room, where the temperature was kept as low as −20°C. In the meantime, the conversion rate of the formed hydrate powder was measured. If the conversion rate reached over 60%, then the hydrate powder was ready to be inserted into the TPCP for NGH pelletizing.

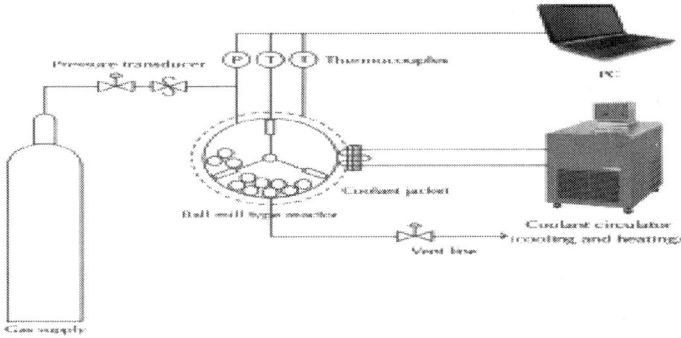

Figure 1: Schematic diagram of the production of NGH.

Hydrate Pellet Extrusion

In the process of pelletizing through the TPCP system, three major design parameters that might affect the mechanical strength of the NGH pellets after production were considered in this study. They were the pressure ratio (PR), feeding pressure (FP), and rotational speed of the twin roll (RPM). We investigated the influence of those three design parameters using two different conditions, respectively. Here, the pressure ratio is defined as the ratio of the output area to the input area of the feeder and is directly related to the level of the feeding quantity of hydrate powder to the TPCP. Also, the feeding pressure can be simply expressed by the vertical weight applied to the top of the feeder. The FP basically indicates the amount of force for pushing the powder into the TPCP. From the preparatory experiments, two experimental conditions for each PR, FR, and RPM are determined. All the experimental conditions, that is, eight cases of different design parameters, are depicted in Table 1.

Table 1: Case of pelletizing using TPCP

Case	1	2	3	4	5	6	7	8
PR	1:2	1:2	1:2	1:2	1:3	1:3	1:3	1:3
FP	2.5 kgf	5 kgf	2.5 kgf	5 kgf	2.5 kgf	5 kgf	2.5 kgf	5 kgf
RPM	1	1	2	2	1	1	2	2

Figure 2 presents the schematics of the proposed twin-roll press for the continuous pelletizing (TPCP) system. It appears that the productivity of the NGH pellet depends on the size of the die hole and the rotational speed of the twin rolls. The diameter of each disk of the twin roll was 750 mm, and the width was 10 mm. The distance between two rollers was fixed as 10 mm. The pellet strip of NGH extruded from the TPCP was cut into rectangular parallelepipeds of 10 mm × 10 mm × 11 mm dimensions. The NGH pellet strips were used for the compressive strength test. For each case in Table 1, ten specimens were equally prepared for the test. Note that all the aforementioned processes were conducted inside the −20°C freezer, to prevent dissociation of the NGH. Figure 3 illustrates the laboratory-scale twin-roll press machine for the continuous pelletizing system in the temperature-controlled chamber.

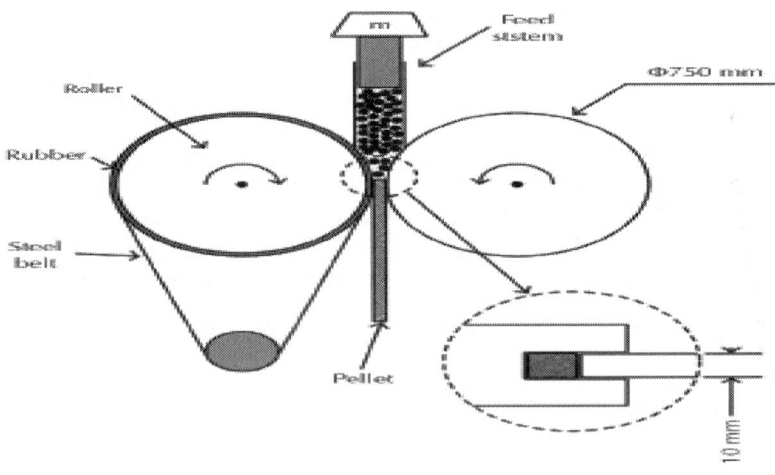

Figure 2: Schematics of the twin-roll press machine for continuous pelletizing.

Figure 3: The lab-scale twin-roll press machine for continuous pelletizing.

Compressive Test of the Hydrate Pellet

In order to assess the mechanical strength of the NGH pellets, a compressive test was conducted using a universal test machine (UTM). The UTM has an environmental chamber that can control the test temperature, ranging from −40°C up to 100°C. In this study, all the compressive tests were conducted at temperature below −25°C. The NGH pellet specimen was placed between two horizontal disks and then slowly pressed by motor-driven UTM, until it developed cracks and eventually collapsed. The load cell on the bottom plate disk recorded the compressive load versus displacement data, to evaluate the compressive strength of each test specimen. Figure 4 illustrates the settings of the compressive test apparatus.

(a)

(b)

Figure 4: Compressive strength test of NGH pellet specimen in UTM.

RESULTS AND DISCUSSION

Mechanical properties, particularly the compressive strength of the NGH pellet, become some of the most important data, in designing the transportation system of solid-form natural gas. Here, the compressive strengths of NGH pellets under different pelletizing conditions of TPCP system were investigated; two PRs were 1:2 and 1:3, two FPs were 2.5 kgf, and 5 kgf, two RPMs were 1 rpm and 2 rpm, respectively. Note that each of three design parameters having two conditions generates in total eight different extrusion conditions for NGH pellet production, as shown in Table 1.

The load-displacement curve of NGH pellet shows the typical pattern of elastoplastic response. Figure 5 shows a load-displacement curve of NGH pellet extruded from TPCP under the test condition of case 6 (PR 1:3, FP 5 kgf, and RPM 1). The curve depicts two moduli, that is, elastic (E_e) and plastic modulus (E_p). In this curve, the compressive strength of the NGH pellet is defined as the intersecting point of the linear curve fitting line before and after the occurrence of the yielding phenomenon. Note that the compressive strength is defined as a yielding load. The slope of the linear curve fitting line before the yielding point becomes the compressive stiffness of the NGH pellet.

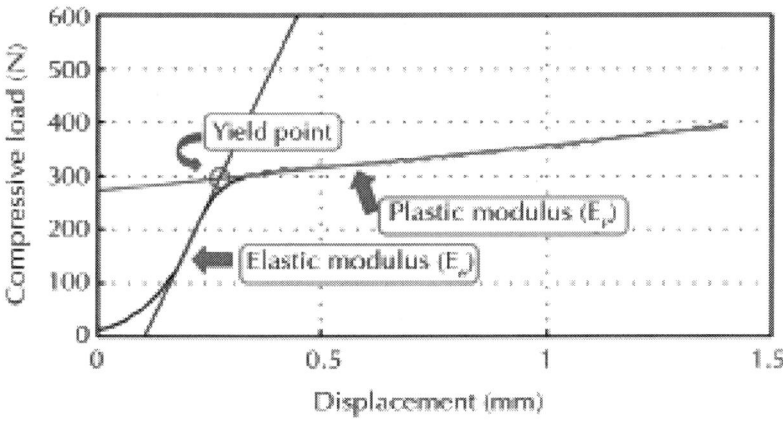

Figure 5: Load-displacement curve of NGH pellet (PR 1:3, FP 5 kgf, RPM 1).

Figure 6 illustrates the load-displacement curves of NGH pellet specimens fabricated under the conditions of cases 1, 2, 3, and 4. In test case 1 through 4, all pressure ratios were fixed to 1:2, but feeding pressure and rotational speed were different for each case. The compressive strength and stiffness of the hydrate pellets were enhanced as the feeding pressure and rotational speed increased. Ten specimens were tested for each case, and average values of compressive strength and stiffness are summarized in Table 2. For case 1 and case 2, hydrate pellets were extruded under the same pressure ratio (1:2) and rotational speed (1 rpm), but different feeding pressure. The yielding loads of hydrate pellets were 341.2 N and 545.5 N for cases 1 and 2, respectively. In this case, the yielding load of NGH pellet increased up to 59.9% as the feeding pressure doubled. Cases 3 and 4 exhibited similar trends, and yielding load increased 54.2%. It is observed that the yielding load of NGH pellet extruded under the same condition of pressure ratio and rotational speed was enhanced, as the feeding pressure increased. From another point of view, cases 1 and 3 provided the same pressure ratio and feeding pressure condition, but different rotational speed of the TPCP system. In this case, the yielding load of NGH pellet increased 43.1% from case 1 through 3, which represents that yielding loads were proportional to the rotational speeds. The same phenomenon was observed when we examined cases 2 and 4. The experimental result shows that the yielding load of NGH pellets was proportional to the feeding pressure and rotational speed, under the constant pressure condition. Therefore, case 4 provided the largest yielding load among the four different cases. Also, the stiffness of NGH pellet showed the same trend.

Table 2: Average yielding load and stiffness of NGH pellets

Unit: yielding load (N), stiffness (N/mm)								
Case	1	2	3	4	5	6	7	8
Yielding load	341.2	545.5	488.2	752.9	168.9	294.3	439.9	495.5
Stiffness	1578.5	1640.5	2793.3	3121.3	998.3	1543.3	1587.8	1891.5

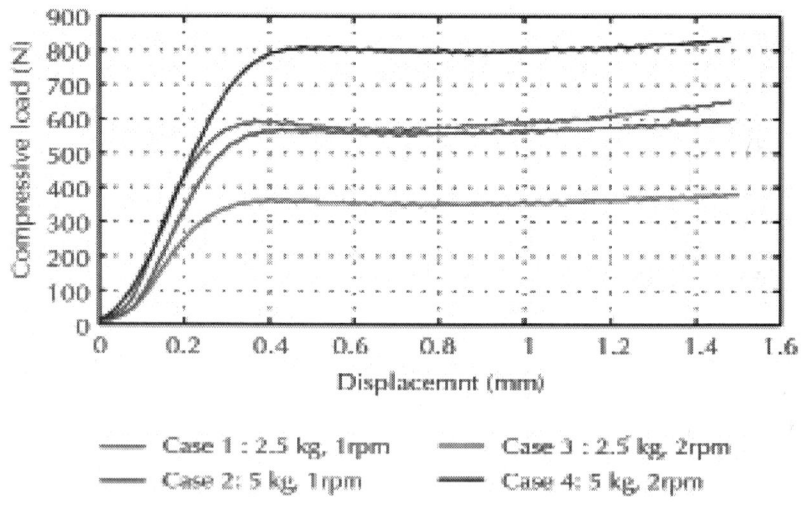

Figure 6: Load-displacement curve of NGH pellet (cases 1, 2, 3, and 4).

Figure 7 illustrates the load-displacement curves of NGH pellets fabricated under the conditions of cases 5, 6, 7, and 8. In cases 5 through 8, the pressure ratio was fixed to 1 : 3, but feeding pressure and rotational speed were different for each case. For cases 5 and 6, NGH pellets were extruded under the same pressure ratio (1 : 3) and rotational speed (1 rpm), but different feeding pressure. The yielding loads of hydrate pellets were 168.9 N and 294.3 N for cases 5 and 6, respectively. Here, the yielding load of NGH pellet increased up to 74.2%, if the feeding pressure was doubled. Cases 7 and 8 provide the same trend, and the yielding load increased 4.5%. It is also noticed that the yielding load of NGH pellet extruded under the same pressure ratio and rotational speed improved, as the feeding pressure increased, for the pressure ratio of 1 : 3. In the sense of rotational speed of cases 5 through 8, the yielding load increased proportional to the rotational speed, which was observed in cases 1 through 4. The only different test parameter between cases 1~4 and cases 5~8 was pressure ratio. The yielding load and stiffness of NGH pellet for relatively low pressure ratio (1 : 2) were higher than those of high pressure ratio (1 : 3) conditions. The pressure ratio has a strong relationship to the amount of feeding NGH powder. Therefore, it is concluded that the higher the pressure ratio condition feed, the less the amount of NGH powder to the die hole, which results in the lowered yielding load. The yielding load and stiffness of hydrate pellets extruded under the eight different conditions are summarized in Figure 8. The yielding load and stiffness of hydrate pellet were enhanced as the feeding pressure and rotational speed increased. The same effect also occurred as the pressure ratio decreased.

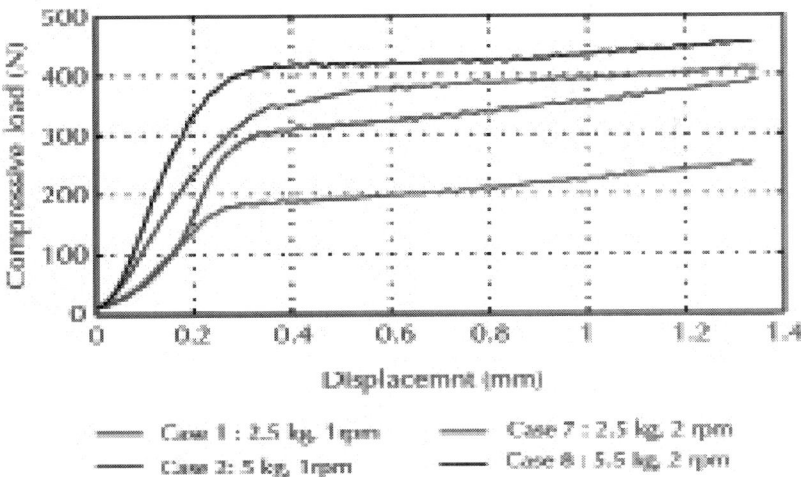

Figure 7: Load-displacement curve of NGH pellet (cases 5, 6, 7, and 8).

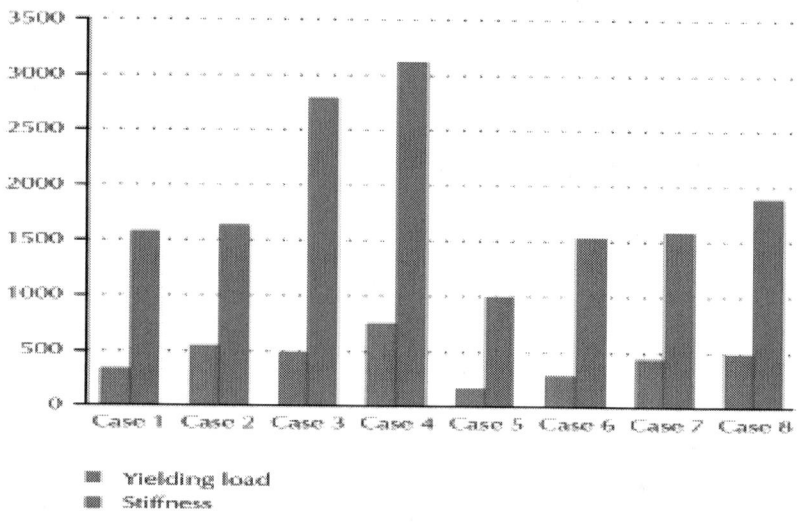

Figure 8: Yielding load and stiffness of NGH pellet.

The yielding load and stiffness of NGH pellet are closely related to its mass density. The mass densities of NGH pellets extruded under the eight different conditions are listed in Table 3. The trend of mass density shows a similar pattern to the yielding load and stiffness. From these observations, the best condition for the largest yielding load and stiffness of hydrate pellet is case 4, which has a feeding pressure of 5 kgf, rotational speed of 2 rpm, and pressure ratio of 1 : 2. It is also concluded that the compressive strength of the NGH pellet is significantly affected by the feeding pressure, rotational speed, and pressure ratio, which constitute fundamental design conditions for the mass-production system of NGH pellets.

Table 3: Mass densities of NGH specimens

Unit: kg/m³								
Case	1	2	3	4	5	6	7	8
Mass density	697.7	750.0	706.8	804.6	686.4	720.5	777.3	786.4

CONCLUSIONS

In this study, the compressive strength of NGH pellet extruded by a twin-roll press for continuous pelletizing system is investigated. The NGH powder was first produced by a laboratory-scale hydrate reactor. NGH pellet strip was continuously extruded by the newly developed TPCP system. The feeding pressure, rotational speed, and pressure ratio were considered as the primary test conditions for making NGH pellet. Two different feeding pressures, rotational speeds, and pressure ratios were investigated, and hydrate pellets were extruded, according to eight different fabrication conditions. The yielding load, stiffness, and mass density of hydrate pellets were highly affected by these three design parameters, and they increased, as the feeding pressure and rotational speed increased. The same effect could also be achieved by decreasing the pressure ratio. Maximum yielding load and stiffness are 752.9 N and 3121.3 N/mm at PR 1 : 2, FP 5 kgf, and 2 rpm, respectively. These values are thought to be enough compressive strength for solid transportation of NGH. The experimental data could be used for solving the design problems of solid form of a gas or NGH pellet transporting system in the near future.

ACKNOWLEDGMENT

This work was supported by the Korea Research Foundation Grant funded by the Korean Government (MOEHRD) (2006RER03P010000), in which main calculations were performed by using the supercomputing resource of the Korea Institute of Science and Technology Information (KISTI).

REFERENCES

1.   J. S. Gudmundsson, "Method for production of gas hydrates for transportation and storage," U.S. patent no. 5536893A, July 1996.

2.   S. H. Lee, Y. S. Yoon, and K. J. Seong, "Experimental study on the dissociation characteristics of methane hydrate pellet by hot

water injection," Transactions of the Korean Society of Mechanical Engineers, B, vol. 35, no. 11, pp. 1177–1184, 2011.

3.   S. Watanabe, S. Takahashi, and H. Mizubayashi, "A demonstration project of NGH land transportation system," in Proceedings of the 4th International Conference on Gas Hydrates, Yocohama, Japan, May 2008.

4.   D. W. Davidson, S. K. Garg, S. R. Gough et al., "Laboratory analysis of a naturally occurring gas hydrate from sediment of the Gulf of Mexico," Geochimica et Cosmochimica Acta, vol. 50, no. 4, pp. 619–623, 1986.

5.   J. S. Gudmundsson and O. F. Graff, "Hydrate non-pipeline technology for transport of natural gas," inProceedings of the 22nd World Gas Conference, 2003.

6.   H. Kanda, "Economic study on natural gas transportation with natural gas hydrate (NGH) pellets," inProceedings of the 23rd World Gas Conference, pp. 1990–2000, June 2006.

7.   H. J. Hong and M. H. Song, "Decomposition characteristics of methane hydrate powder during decompression," in Proceedings of KSME Annual Conference, pp. 13–17, Seoul, Republic of Korea, January 2010.

8.   T. Iwasaki, Y. Katoh, S. Nagamori, S. Takahashi, and N. Oya, "Continuous natural gas hydrate pellet production (NGHP) by process development unit (PDU)," in Proceedings of the 5th International Conference on Gas Hydrates, 2005.

9.   J.-H. Choi and B.-H. Koh, "Compressive strength of ice-powder pellets as portable media of gas hydrate," International Journal of Precision Engineering and Manufacturing, vol. 10, no. 5, pp. 85–88, 2009.

10.   T. J. Ko, J. H. Kim, and I. J. Yoon, "A study of WC end-milling manufacturing and cutting ability evaluation by using powder injection molding," International Journal of Precision Engineering and Manufacturing, vol. 10, no. 4, pp. 13–17, 2009.

11.   H. B. Shim, "Improving formability to develop miniature stamping technologies," International Journal of Precision Engineering and Manufacturing, vol. 10, no. 2, pp. 117–125, 2009.

12. U. Jung, J. H. An, B. S. Lim, and B. H. Koh, "Modeling discharge of pellets from a hopper using response surface methodology," International Journal of Precision Engineering and Manufacturing, vol. 13, no. 4, pp. 565–571, 2012.

Chapter 9

Phenolic Extractives and Natural Resistance of Wood

[1]Departamento de Antibióticos, Universidade Federal de Pernambuco, UFPE, Recife, PE, Brazil, Brazil

[2]Departamento de Química, Universidade Federal Rural de Pernambuco, UAST, Serra Talhada, PE, Brazil

[3]Instituto Federal Educação, Ciência e Tecnologia de Pernambuco, IFPE, Campus Recife, Recife, PE, Brazil

[4]Departamento de Química Fundamental, Universidade Federal de Pernambuco, UFPE, Recife, PE, Brazil, Brazil

INTRODUCTION

Wood is a natural organic material that consists mainly of two groups of organic compounds: carbohydrates (hemicelluloses and cellulose) and phenols (lignin), that correspond to (65-75%) and (20-30%), respectively (Pettersen 1984). The wood is also constituted of minor amounts of extraneous materials, mostly in the form of organic extractives (usually 4–10%) and inorganic minerals (ash), mainly calcium, potassium, and magnesium, besides manganese and silica.

Generally, wood has an elemental composition of about 50% carbon, 6% hydrogen, 43% oxygen, trace amounts of nitrogen and several metal ions.

Cellulose is a long-chain linear polymer exclusively constructed of β-1,4-linked D-glucose units which can appear as a highly crystalline material (Fan et al, 1982). Often 5000 to 15000 glucose rings are polymerized into a single cellulose molecule.

Hemicelluloses consist of relatively short heteropolymers consisting of the pentoses D-xylose and L-arabinose and the hexoses, D-glucose, D-mannose, D-galactose, D-rhamnose and their corresponding uronic acids. It is composed of only 500-3000 sugar units, and thus has a shorter chain than cellulose (Saka 1991)

Lignin, the third cell wall component, is an aromatic polymer synthesized from phenylpropanoid precursors (Adler 1977). It is a three-dimensional polymer formed of coniferyl, syringyl, and coumaryl alcohol units with many different types of linkages between the building blocks and by far the most complex of all natural polymers.

Extractives are chemical constituents residing in the lignocellulosic tissue that contains an higher diversity of organic compounds, for example triglycerides, steryl esters, fatty acids, sterols, neutral compounds, such as fatty alcohols, sterols, phenolic compounds such as tannins (Fava et al, 2006), quinones (Carter et al, 1978; Ganapaty et al, 2004), flavonoids (Reyes-Chilpa et al, 1995; Ohmura et al, 2000; Chen et al, 2004; Morimoto et al, 2006; Sirmah et al, 2009), besides terpenoids (Kawaguchi et al, 1989; Chang et al, 2000; Watanabe et al, 2005) and alkaloids (Kawaguchi et al, 1989).

EXTRACTIVES AND NATURAL RESISTANCE OF WOOD

Cellulose is the major structural component of wood and also the major food of insects and decay fungi. Termites, like fungi, are important biological agents in the biodegradation of wood (Syofuna et al, 2012).

Extractives are low molecular weight compounds present in wood (Chang et al, 2001), also called secondary metabolites, and are indeed crucial for many important functional aspects of plant life. The relationship between extractives and natural durability of wood was first reported by Hawley et al (1924). The natural durability of wood is often related with its toxic extractive components (Scheffer and Cowling 1966; Carter et al, 1978; Hillis 1987; McDaniel 1992; Taylor 2006; Santana et al, 2010).

Heartwood extractives retard wood decay can protect the wood against decay organisms (Walker 1993,Hinterstoisser et al, 2000; Schultz and Nicholas 2002), but the natural durability is extremely complex and additional factors such as density of wood and lignin content, besides this dual fungicidal and antioxidant action, may be involved (Schultz and Nicholas 2002).

Several studies have shown that after removal of extractives, durable wood loses its natural resistance and makes them more susceptible to decay (Ohmura, 2000; Taylor et al, 2002; Oliveira et al, 2010). Several authors investigated the relationships between the wood properties and extractives (Carter et al, 1978; Schultz et al, 1990; Reyes-Chilpa et al, 1998; Chang et al, 1999; Morimoto et al, 2006).

One of the most limiting factors for the commercial utilization of wood is its low resistance to fungi and termites, especially in the semi-arid and sub-humid tropics. The biodegradation is supposed to be one of the major challenges to incur the heavy economic loss. Wood decay fungi and some species of termites are important and potent wood-destroying organisms attacking various components of the wood (Istek et al, 2005; Gonçalves and Oliveira 2006).

The largest group of fungi that degrades wood is the basidiomycetes and is divided into: white-rot, brown- rot and soft-rot fungi (Anke et al, 2006). Brown-rot fungi occurs most often in buildings, can degrade only structural carbohydrates (cellulose and hemicellulose), leaving lignin essentially undigested, whereas white-rot fungi utilize all wood constituents including both the carbohydrates and the lignin. Soft-rot fungi utilize preferably carbohydrates, but also degrade lignin (Belie et al, 2000). They hydrolyze and assimilate as food the lignocellulose components by injecting enzymes into the wood cells (Erickson et al, 1990).

Termites cause significant losses to annual and perennial crops and damage to wooden components in buildings (Verma 2010). Damage caused by subterranean termites, Nasutitermes, Coptotermes and Reticulitermes historically has been a concern of researchers worldwide. Korb (2007) estimated annual damage caused by termites at about U.S. $50 billion worldwide. In the city of Sao Paulo, Brazil, alone, a 20-year loss of $3.5 billion was incurred (Lelis, 1994).

The concentration of extractives varies among species, between individual trees of the same species and within a single tree. Some of these extractives render the heartwood unpalatable to wood destroying organisms. Factors affecting wood consumption by termites and fungi are numerous and complexly related. The amount however can vary from season to season even in the same tissue or are restricted in certain wood species (Taylor et al, 2006).

Several woods contain extractives which are toxic or deterrent for termites, bacteria and fungi resistance (Maranhão 2013; Taylor et al, 2006). Termite resistance of wood is a function of heartwood extractive variability while individual extractives inhibit fungal growth (Neya at al, 2004; Arango et al, 2006).

Biological deterioration of wood is of concern to the timber industry due to the economic losses caused to wood in service or in storage. Fungi, insects, termites, marine borers and bacteria are the principal wood biodegraders. They attack different components of wood at different rates giving rise to a particular pattern of damage (Sirmah 2009). Degradation is influenced by environmental conditions of the wood; whether in storage or in use. The degraded wood material is returned into the soil to enhance its fertility (Silva et al, 2007).

The proposal of this study is to demonstrate the importance of phenolic compounds in natural resistence of wood biodegradation. We collected information of the most representative phenolic compounds (flavonoids, stilbenes, quinones and tannins) found in wood, responsible for resistance of some wood species to bio-degraders (Toshiaki 2001; Windeisen et al, 2002).

FLAVONOIDS

Flavonoids are secondary metabolites that occur naturally in all plant families (Harbone 1973). Widely distributed in all parts of plants, these compounds afford protection against ultraviolet radiation, pathogens, and herbivores (Harbone and Willians 2000). The general structure includes a C15 (C_6-C_3-C_6) skeleton joined to a chroman ring (benzopyran moiety), classified into flavanones, flavones, chalcones, dihydroflavonols, flavonols, aurones, flavan-3-ols, flavan-3,4-diols, anthocyanidins, isoflavonoids, and neoflavonoids. Some examples of each class of flavonoids are described in figure 1.

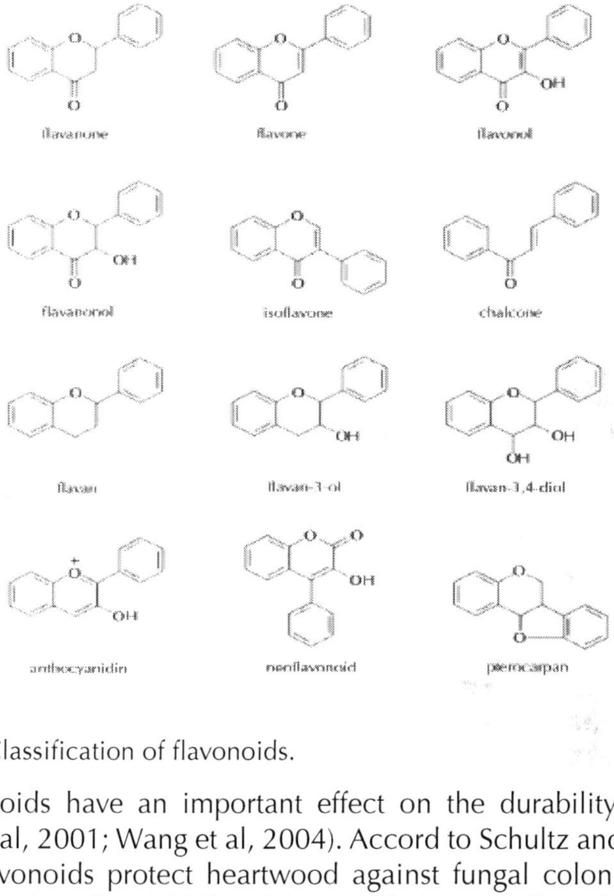

Figure-1: Classification of flavonoids.

Flavonoids have an important effect on the durability of wood (Chang et al, 2001; Wang et al, 2004). Accord to Schultz and Nicholas (2000) flavonoids protect heartwood against fungal colonization by a dual function: fungicidal activity and being excellent free radical scavengers (antioxidants). Flavonoids are natural antioxidants and have received attention due to their role in the neutralization or scavenging of free radicals (Gupta and Prakash 2009). Pietarinen (2006) showed that the radical scavenging activity is particularly important because both white-rot and brown-rot fungi are believed to use radicals to disrupt cell walls.

The heartwood of Lonchocarpus castilloi Standley (Leguminosae) is highly resistant to attack by the dry wood termites Cryptotermes brevis (Walker) (Isoptera: Kalotermitidae). Two flavonoid isolated from the heartwood of this plant, castillen D and castillen E (Figure 2), that presented feeding deterrent activity to C. brevis (Reves-Chilpa et al, 1995).

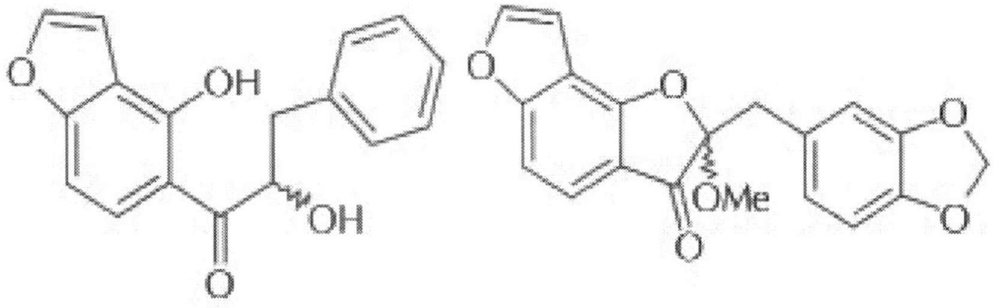

castillen E

castillen D

Figure-2: Structure of castillen E and castillen D.

Ohmura et al (2000) reported that flavonoids present in Larix leptolepis (Pinaceae) wood, principally taxifolin and aromadedrin, showed strong feeding deterrent activities against the subterranean termite,Coptotermes formosanus Shiraki (Isoptera: Rhinotermitidae) and suggested that some flavonoids such as quercetin and taxifolin (Figure 3) might be useful for termite control agents considering their abundance in plants.

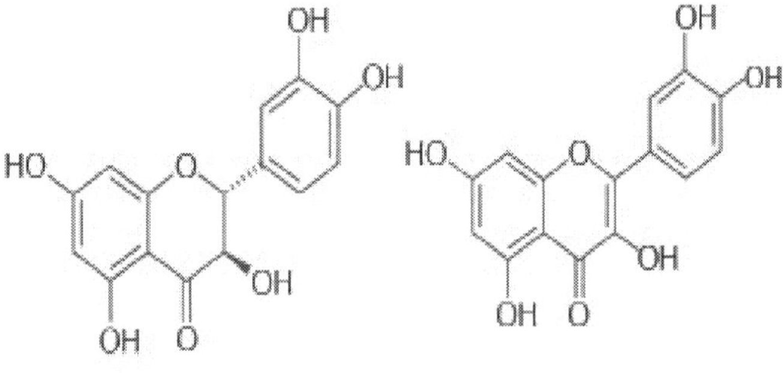

taxifolin

quercetin

Figure-3: Structure of quercetin and taxifolin.

The heartwood of Acacia auriculiformis (Leguminosae) has been shown to contain a number of different flavonoids and proanthocyanidins content (Sarai et al, 1980; Barry et al, 2005). According to Schultz et al (1995) the durability of Acacia species was attributed the presence of dihydromorin and aromadedrin (Figure 4).

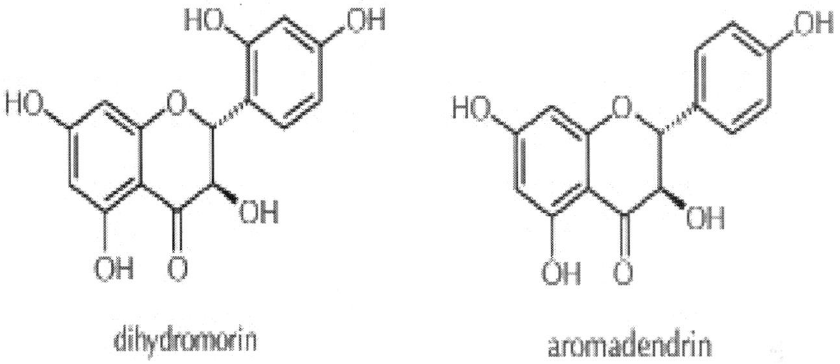

dihydromorin   aromadendrin

Figure-4: Structure of dihydromorin and aromadendrin.

From heartwood of Morus mesozygia (Moraceae), besides dihydromorin, were isolated morin and pinobanksin (Figure 5), but the resistance against wood destroying basidiomycetes, Coriolus versicolor, Lentinus squarrosulus and Poria spp. was related to the presence of dihydromorin (Toirambe Bamoninga and Ouattara, 2008).

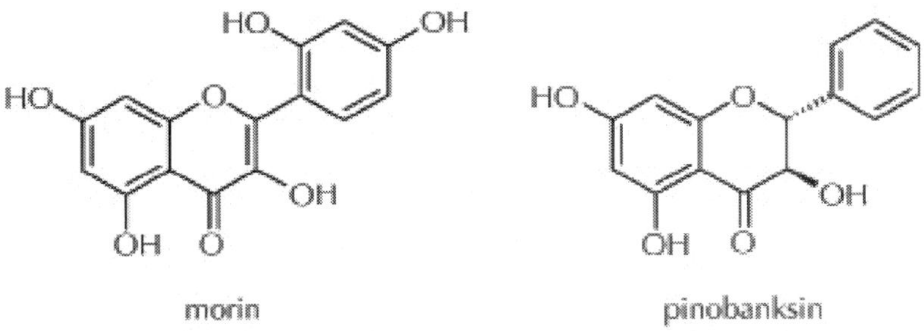

morin   pinobanksin

Figure-5: Structure of morin and pinobanksin.

According to Sirmah et al (2009) the durability of Prosopis juliflora wood (Leguminosae) was assigned to (−)-mesquitol (Figure 6), but

Pizzo et al (2011) related that (-)-mesquitol alone cannot be considered the single most important factor in determining the durability of the Prosopis species. Laboratory tests indicated that the heartwood of P. juliflora was resistance against to both white- and brown-rot fungi (Sirmah 2009).

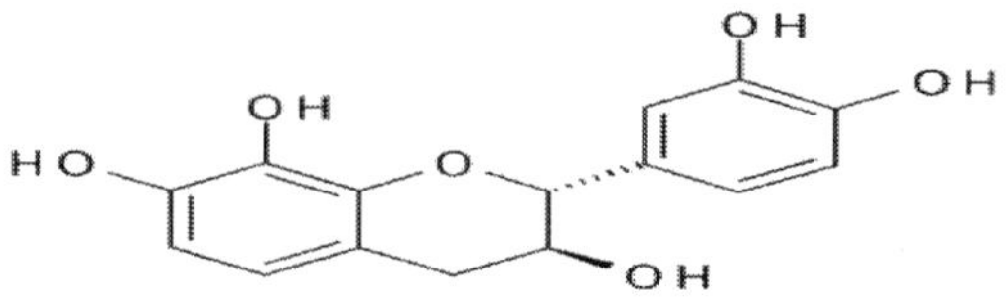

(-)-mesquitol

Figure-6: Structure of mesquitol.

The antifeedant activity of some flavonoids against the subterranean termite Coptotermes formosanusShiraki was examined with no-choice tests and two-choice tests (Ohmura et al, 2000). The structure-activity relationships of these flavonoids (Figure 7) were evaluated and it was found that flavonoids containing hydroxyl groups at C-5 and C-7 in A-rings showed high antifeedant activity. Furthermore, the presence of a carbonyl group at C-4 in the pyran rings of the compounds was necessary for the occurrence of high activity. 3-hydroxyflavones and 3-hydroxyflavanones with 3', 4'- dihydroxylated B-rings exhibited higher activity than those with 4'-hydroxylated B-rings.

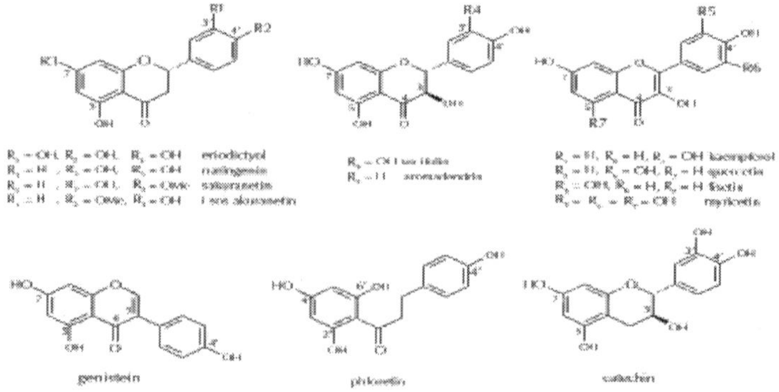

Figure-7: Flavonoids and antifeedant activity against the subterranean termite C. formosanus.

The antifeedant activities of pterocarpans isolated from the heartwood of Pterocarpus macrocarpusKruz. (Leguminosae) were evaluated against the subterranean termite, Reticulitermes speratus Kolbe (Isoptera: Rhinotermitidae). Three isolated pterocarpans, (-)-homopterocarpin, (-)-pterocarpin, and (-)-hydroxyhomopterocarpin were tested (Figure 8). The most active antifeedant against R. speratus was (-)-homopterocarpin. However, all pterocarpans showed antifeedant activity against R. speratus(Morimoto et al, 2006).

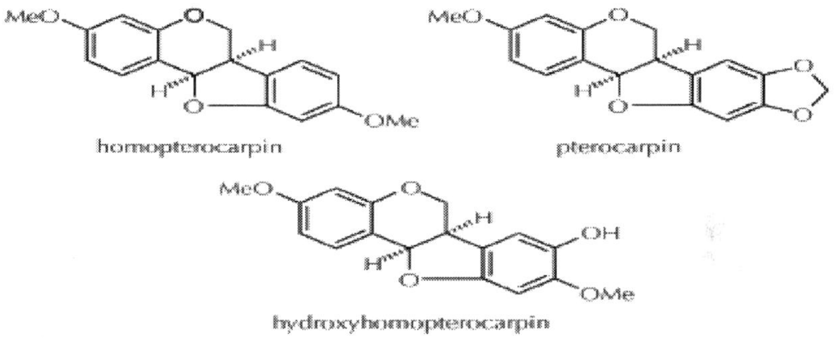

homopterocarpin

pterocarpin

hydroxyhomopterocarpin

Figure-8: Structure of homopterocarpin, pterocarpin and hydroxyhomopterocarpin.

From the heartwood of Dalbergia latifolia (Leguminosae) were isolated and identified as active against termites and fungi, the neoflavonoids, latifolin, dalbergiphenol, and 4-methoxydalbergione (Figure 9).

latifolin

dalbergiphenol

4-methoxydalbergione

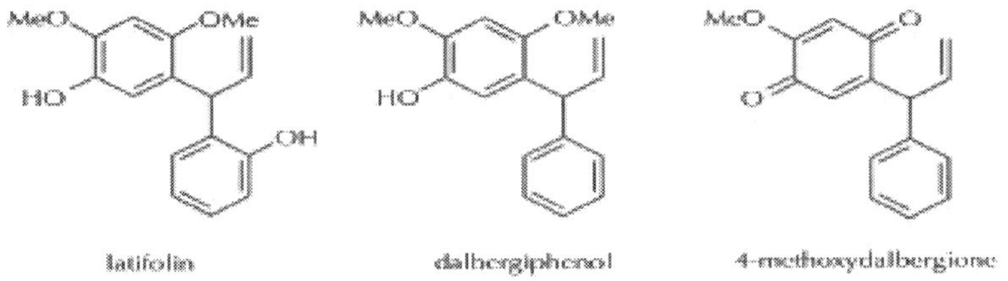

Figure-9: Structure of latifolin, dalbergiphenol, and 4-methoxydalbergione.

With respect to activity against Trametes versicolor, a white-rot basidiomycete, latifolin and 4-methoxydalbergione showed activity.

Dalbergiphenol exhibited relatively high antifungal activity against the brown-rot basidiomycete, Fomitopusis palustris (Sekine et al, 2009).

Latifolin showed high termiticidal activity and termite-antifeedant against Reticulitermes speratus(Kolbe). Dalbergiphenol and 4-methoxydalbergione exhibited moderate termite-antifeedant activity (Sekine et al, 2009).

The structure-activity relationships of latifolin (Figure 10) and its derivatives were analyzed to check if there was a correlation between antitermitic and antifungal activity. It was found that the termite mortality in response to the derivatives 2′-O-methyllatifolin, latifolin dimethyl ether, and latifolin diacetate increased 2-fold compared to latifolin. No difference was presented in mortality of termites in the presence of 5-O-methyllatifolin and latifolin. The results indicate that the phenolic hydroxyl group at C-5 of the A ring provides antitermitic activities.

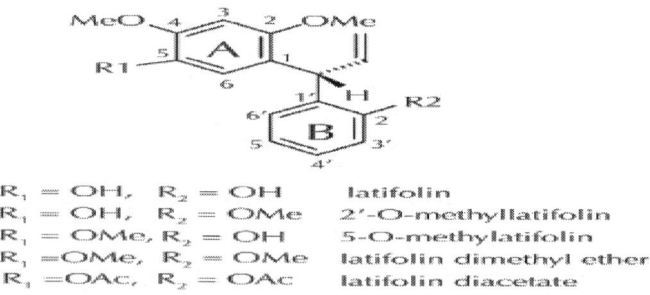

R_1 = OH, R_2 = OH   latifolin
R_1 = OH, R_2 = OMe   2′-O-methyllatifolin
R_1 = OMe, R_2 = OH   5-O-methylatifolin
R_1 = OMe, R_2 = OMe   latifolin dimethyl ether
R_1 = OAc, R_2 = OAc   latifolin diacetate

Figure 10: Structure of latifolin and its derivatives.

With respect to antifungal activity of these compounds, it was found that all compounds presented less activity against white- and brown-rot fungi than latifolin. In addition, both C-5 and C-2′ phenolic hydroxyl groups in the A and B rings have antifungal activity against white- and brown-rot fungi. In conclusion, the bioactivity of latifolin depends upon the position of phenolic hydroxyl groups (Sekine et al, 2009).

The heartwood of Dalbergia congestiflora Pittier (Leguminosae) tree presented natural resistance to fungal attack. The antifungal effect of various extracts from the D. congestiflora heartwood was evaluated against Trametes versicolor fungus (Martínez-Sotres et al, 2012).

The major component of hexane extract that caused 100% growth inhibition from tested fungi was characterized as (-)-Medicarpin (Figure 11). Medicarpin also isolated from heartwood of Platymiscium yucatanum(Leguminosae) was identified active against T. versicolor (Reyes-Chilpa et al, 1998).

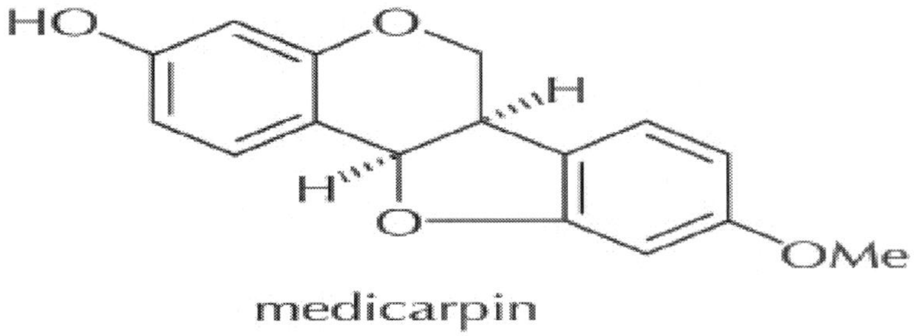

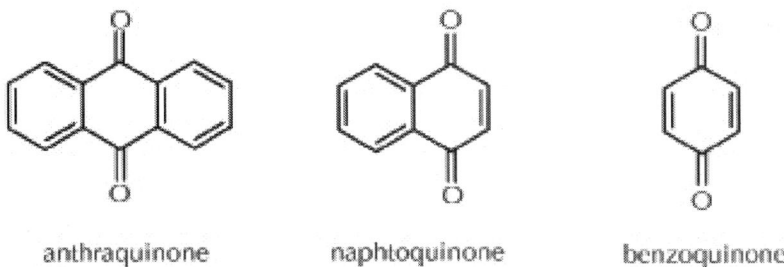

Figure-11: Structure of medicarpin.

QUINONES

Various types of quinones (benzoquinones, naphthoquinones, or anthraquinones) occur in many plant families (Toshiaki 2001). The above mentioned classification of quinones is described in Figure 12. Termite resistant woods are said to contain allelochemicals such as quinones that possess natural repellent and toxic properties (Carter et al, 1978; Scheffrahn 1991; Ganapy et al, 2004; Dungani et al, 2012).

anthraquinone   naphtoquinone   benzoquinone

Figure-12: Classification of quinones.

The heartwood of Tectona grandis L. f. (Lamiaceae) contains a large amount of quinones that possess considerable influence on the natural durability of teak wood. The naphthoquinone, 4', 5'–dihydroxyepiisocatalponol (Figure 13) plays a key role in the resistance of teak against fungi attack. In-vitro bioassays indicated that this compound acted as a fungicide against the White-rot fungi Trametes versicolor (Niamké et al, 2012). Tectoquinone (Figure 13), a anthraquinone, presented strong antitermitic activity and is assumed to be at the origin of the resistance of teak wood to termites (Haupt et al, 2003; Kokutse et al, 2006). According to Wolcott (1955) this substance is highly repellent to the dry-wood termite Cryptotermes brevis (Walker) and Sandermann and Dietrichs (1957) demonstrated its toxicity to subterranean termite Reticulitermes flavipes.

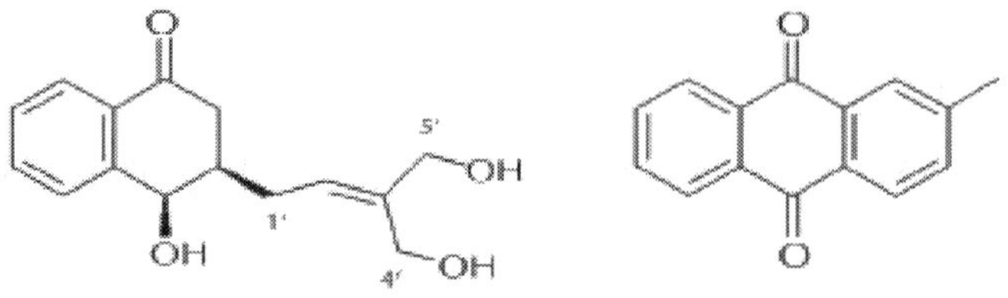

4',5'-dinydroxy-epiisocatalponol   tectoquinone

Figure-13: Structure of 4', 5'–dihydroxyepiisocatalponol and tectoquinone.

Castillo and Rossini (2010) isolated naphthoquinones from heartwood of Catalpa bignonioides(Bignoniaceae) that showed activity against the termite Reticulitermes flavipes. The most abundant and active termiticidal compounds were catalponol and catalponone (Figure 14).

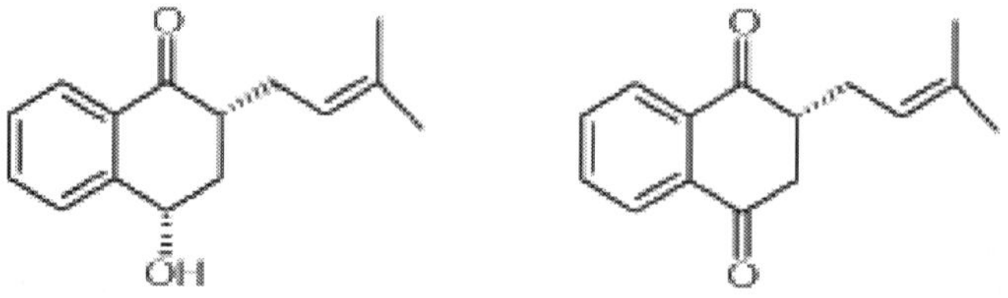

catalponol   catalponone

Figure-14: Structure of catalponol and catalponone.

From heartwood of Tabebuia impetiginosa (Bignoniaceae) were isolated naphthoquinones, mainly lapachol (Figure 15), that showed no repellent activity to Reticulitermes termites but it was repellent to two other termites, Microcerotermes crassus (Isoptera: Termitidae) and Kalotermes flavicollis(Isoptera: Kalotermitidae) (Becker et al, 1972).

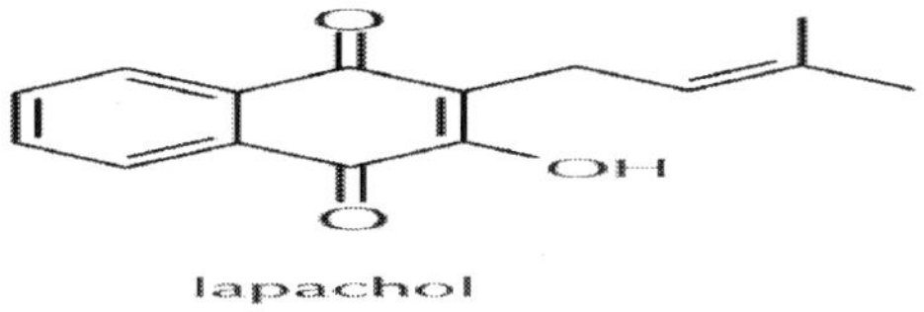

lapachol

Figure-15: Structure of lapachol.

The naphthoquinone, 7-methyljuglone (Figure 16) was isolated and identified as termicidal constituent of heartwood of Diospyros virginiana L. (Ebenaceae). Its dimer, isodiospyrin possess also termicidal activity against Reticulirmes flavipes, but to a lesser extent (Carter et al, 1978).

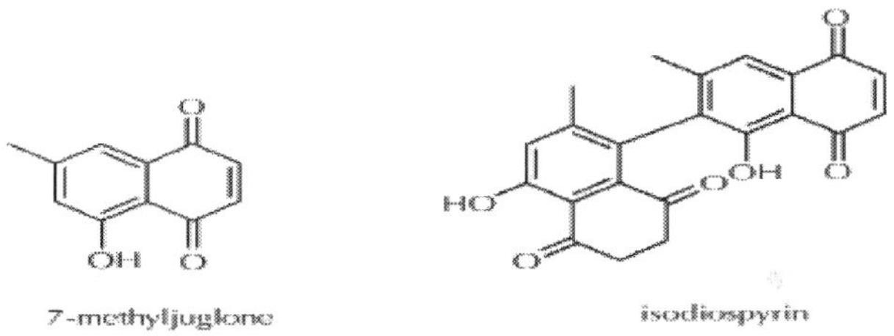

7-methyljuglone   isodiospyrin

Figure-16: Structure of 7-methyljuglone and isodiospyrin.

STILBENES

Stilbenes are compounds possessing the 1,2-diphenylethene structure, as well as bibenzyls and phenanthrenes, which are composed of C_6-C_2-C_6 skeleton. Stilbenes derivatives of 1,2-diphenylethlene, process a conjugated double bond system. There are two isomeric forms of 1,2-diphenylethylene: trans-stilbene and cis-stilbene, and the chemical structure of these two stilbenes are shown in Figure 17.

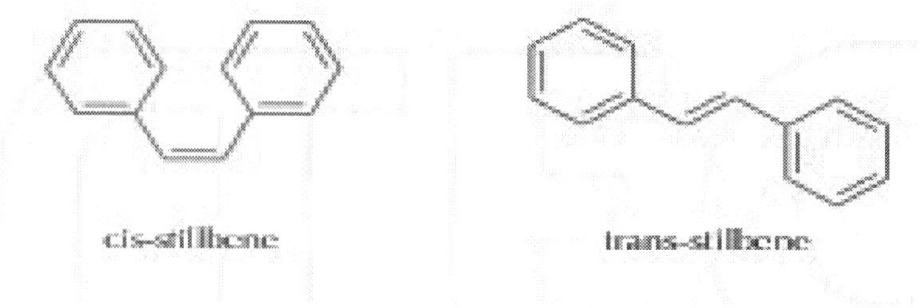

cis-stilbene   trans-stilbene

Figure-17: The chemical structure of stilbenes.

Hydroxylated trans-stilbene has an important role in heartwood durability, especially for a resistance to fungal decay. The durability and resistance to decay by Pinus sylvestris (Pinaceae) is due to pinosylvins (Figure 18). Pinosylvin present in the heartwood of Pinus species is formed as a response to external stress such as fungal infections or UV light. The 2, 4, 3', 5'-tetra and 3, 4, 5, 3', 5'-pentahydroxystilbenes are responsible for wood resistance against Brown-rot and whit-rot fungi (Schultz et al, 1995).

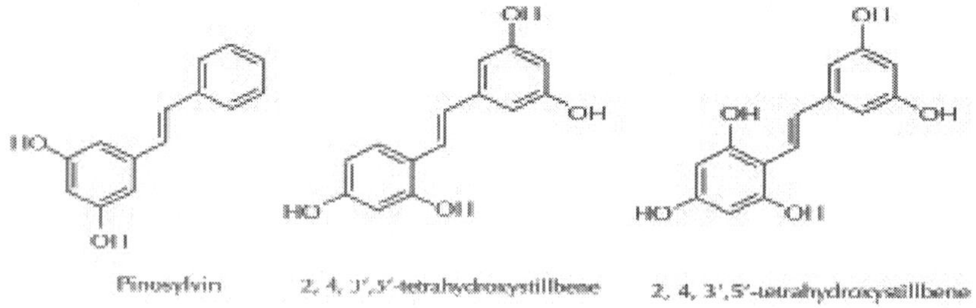

Pinosylvin   2, 4, 3',5'-tetrahydroxystilbene   2, 4, 3',5'-tetrahydroxystilbene

Figure-18: The chemical structure of pinosylvin and derivates.

From the heartwood of Bagassa guianensis (Moraceae) was isolated moracins including others polyphenols such as flavonoids and stilbenoids (Figure 19), that presented activity against Pycnoporus sanguineus, a white-rot fungus. Possible synergism between compounds have been hypothesized (Royer et al, 2012).

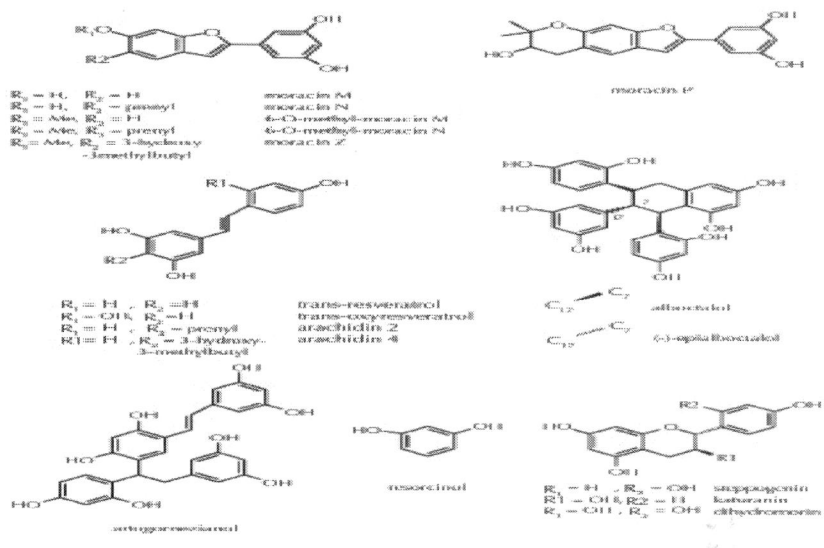

Figure-19: The chemical structure polyphenols from B. guianensis.

TANNINS

Tannins constitute a distinctive and unique group of higher plant metabolites. They presented polyphenolic character and relatively large molecular size (from 500 to >20,000). They are thought by some to constitute one of the most important groups of higher plant defensive secondary metabolites (Haslam 1989).

The designation of tannin includes compounds of two distinct chemical groups: hydrolysable tannins (Figure 20) and condensed tannins (Figure 21).

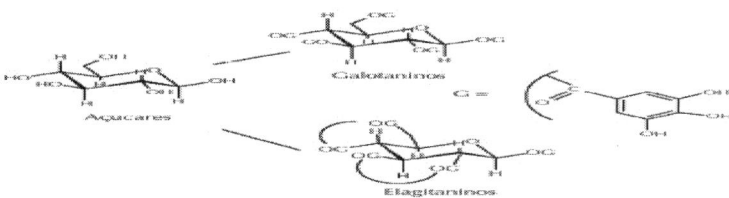

Figure-20: Structure of hydrolysable tannins.

Hydrolysable tannins are molecules with a polyol (D-glucose) as a central core. The hydroxyl groups of these carbohydrates are partially or totally esterified with phenolic groups like gallic acid (gallotannins) or ellagic acid (ellagitannins). Hydrolysable tannins are usually present in low amounts in plants.

Condensed tannins are probably the most ubiquitous of all plant phenolics, and presented exceptional concentrations in the barks and heartwoods of a variety of tree species. They are oligomers or polymers of flavonoid units (flavan-3-ol) linked by carbon-carbon bonds not susceptible to cleavage by hydrolysis (Sirmah 2009).

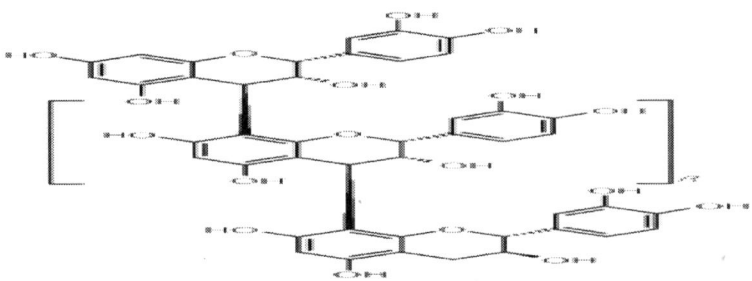

Figure-21: Structure of condensed tannins.

Condensed tannins are natural preservatives and antifungal agents, found in high concentrations in the bark and wood of some tree species (Zucker 1983). Most plant-pathogenic fungi excrete extracellular enzymes such as cellulases and lignases, involved in the invasion and spread of the pathogen. Condensed tannins most likely act as inhibitors of these enzymes by complexing, blocking their action (Peter et al, 2008). For this reason, extract from various woods and barks rich in tannin have been used as adhesives and wood preservatives for a long time (Brandt 1952; Plomely 1966; Mitchell and Sleeter 1980; Pizzi and Merlin 1981; Laks et al, 1988; Lotz and Hollaway 1988; Toussaint 1997; Thevenon 1999).

CONCLUSIONS

The protection of wood against biodeterioration is related to its chemical composition, mainly due to the accumulation of extractives in the heartwood. Wood extractives are nonstructural wood components that play a major role in the susceptibility of wood against wood decay organisms. The attack of these organisms in general can be prevented with synthetic organic and inorganic preservatives; however, such products are very harmful to human health and the environment. Several studies have considered that, it is possible the application of wood extractives as natural preservatives. The main components of wood extractives that confers natural resistance against biodeterioration agents are, tannins, flavonoids, quinones and stilbenes.

- Frequently, condensed tannin can be obtained inexpensively by extracting the bark materials with hot water solvent and has been used as preservatives for a long time.

- Flavonoids exhibit antifungical activity as well as feeding deterrent activities against subterranean termites.

- Quinones – possess natural repellent and toxic properties, mainly against termites.

- Stilbenes has an important role in heartwood durability, especially for a resistance to fungal decay.

The characteristics of all wood species are described in Table 1.

Table-1: List of wood species with their family, common names, resistance and distribution

Scientific name	Familie name	Common name	Resistance	Origin
Acacia auriculiformis	Leguminosae	Australian wattle	Durable wood(Ashaduzzaman et al, 2011)	Australia, Indonesia, Papua New Guinea
Bagassa guianensis	Moraceae	Tatajuba	Very resistant (Rover et al. 2012)	Guianas and Brazil
Catalpa bignonioides	Bignoniaceae	Common CatalpaIndian Bean	Highly decay resistant heartwood (Muñoz-Mingarro et al, 2006)	North America
Dalbergia congestifoliaPittier	Leguminosae	Rosewood	Resistant wood (Martínez-sotres et al, 2012)	Central America
Dalbergia latifolia	Leguminosae	Indian rosewood	Resistant wood (Lemmens, 2008)	Asia
Diospyros virginiana	Ebenaceae	Common persimmon	-	Africa, Asia

Larix leptolepis	Pinaceae	Japanece larch	resistant (Schaffer and Morrell 1998)	Japan
Lonchocarpus castilloi	Leguminosae	Black cabbage bark	very resistant(Schaffer and Morrell 1998)	Latin America
Morus mesozygia	Moraceae	Mulberry	Non-resistant (Schaffer and Morrell 1998)	Africa
Pinus sylvestris	Pinaceae	Redwood, Scots pine	Non-resistant (Schaffer and Morrell 1998)	Europe, Asia
Platymiscium yucatanum	Leguminosae	Granadillo	very resistant (Schaffer and Morrell 1998)	Latin America
Prosopis juliflora	Leguminosae	Mesquite, algarroba	Resistant (Ramos et al, 2006)	South and Central America
Pterocarpus macrocarpus	Leguminosae	Burma padauk	very resistant (Schaffer and Morrell 1998)	Native to Thailand and Myanmar
Tabebuia impetiginosa		Brazil wood	Very resistant (Paes et al, 2005)	Latin America
Tectona grandis L. f.	Lamiaceae	teak	Very resistant (Kokutse et al, 2006)	Native to southern Asia

REFERENCES

1. M Adfa, T Yoshimura, K Komura, M Koketsu, 2010Antitermite activities of coumarin derivatives and scopoletin from Protium javanicum Burm. f. Journal of Chemical Ecology 36720726

2. E Adler, 1977Lignin chemistry-past, present and future. Wood Sci. Technol. 11169218

3. N Amusant, C Moretti, B Richard, E Prost, J. M Nuzillard, M. F Thevenon, 2007Chemical compounds from Eperua falcata and Eperua grandifloraheartwood and their biological activities against wood destroying fungus (Coriolus versicolor). Holz als Roh- und Werkstof 652328

4. H Anke, W Roland, S Weber, 2006White-rots, chlorine and the environment- a tale of many twists. Mycologist2038389

5. Arango, R. A, Green III, F, Hintz, K, Lebow, P. K, & Miller, R. B. (2006). Natural durability of tropical and native wood against

termite damage byReticulitermes flavis (Kollar). International Biodeterioration & Biodegradation, 57, 146-150.

6. M Ashaduzzaman, A. K Das, M. I Shams, 2011Natural Decay Resistance of Acacia auriculiformis Cunn. ex. Benth and Dalbergia sissoo Roxb. Bangladesh J. Sci. Ind. Res., 46225230

7. K. M Barry, R Mihara, N. W Davies, T Mitsunaga, C. L Mohammed, 2005Polyphenols in Acacia mangium and A. auriculiformis heartwood with reference to heart rot. J. Wood Sci. 51615621

8. G Becker, M Lenz, S Dietz, 1972Unterschiede im Verhalten und der Giftempfindlichkeit verschiedener Termiten-Arten gegenuber einigen Kernholzstoffen. Z. Angew. Entomol. 71201214

9. J. R Beckwith, 1998Durability of Wood. University of Georgia School of Forest Resources Extension Publication for 98026

10. N. D Belie, M Richardson, C. R Braam, B Svennerstedt, J. J Lenehan, B Sonck, 2000Durability of Building Materials and Components in the Agricultural Environment: Part I, The agricultural environment and timber structures. J. agric. Engng Res. 75225241

11. T. G Brandt, 1952Mangrove tannin-formaldehyde resins as hot-pressed plywood adhesives. Tectona, 42, 137.

12. F. L Carter, A. M Garlo, J. B Stanely, 1978Termiticidal components of wood extracts: 7-methyl-juglone from Diospyros virginiana. Journal of Agricultural and Food Chemistry 26869873

13. L Castillo, C Rossini, 2010Bignoniaceae Metabolites as Semiochemicals. Molecules 20101570907105

14. S. T Chang, J-H Wang, C. L Wu, P. F Chen, Y. H Kuo, 2000Comparison of the antifungal activity of cadinane skeletal sesquiterpenoid from Taiwania (Tawania Crypromerioides Hayara) heartwood. Holzforschung 543241245

15. S Chang, T Wang, S. -Y Wu, C. -L Su, Y. -C Kuo, Y.-H. 1999Antifungal compounds in the ethyl acetate soluble fraction of the extractives of Taiwania (Taiwania cryptomerioides Hayata) heartwood. Holzforschung 535487490

16. K Chen, W Ohmura, S Doi, M Aoyama, 2004Termite feeding deterrent from Japanese larch wood. Resource Technology 95129134

17. Dungani, R, Bhat, I. U. H, Abdul Khalil, H. P. S, Naif, A, & Hermawan, D. (2012). Evaluation of Antitermitic Activity of Different Extract Obtained from Indonesian Teakwood (Tectona grandis L.f). Journal of Bioresources 7214521461

18. K. E. L Ericksson, R. A Blanchette, P Ander, 1990Microbial and enzymatic degradation of wood and wood components. Springer-Verlag, Berlin, Germany, 407p.

19. L. T Fan, Y. H Lee, M. M Gharpuray, 1982The nature of lignocellulosics and their pretreatments for enzymatic hydrolysis. Adv. Biochem. Eng. 23158187

20. Fava, F, Monteiro de Barros, M, Stumpp, E, & Ramão Marceli Jr., F. (2006). Aqueous extract to repel or exterminate termites. Patent Application WO 2005BR173 20050824.

21. Ganapaty, S, Thomas, P. S, Fotso, S, & Laatsch, H. (2004). Antitermitic quinones from Disopyros sylvatica. Phytochemistry, 65, 1265-1271.

22. F. G Gonçalves, J. T. S Oliveira, 2006Resistência ao ataque de cupim-de-madeira seca (Cryptotermes brevis) em seis espécies florestais. Cerne, 1218083

23. S Gupta, J Prakash, 2009Studies on Indian green leafy vegetables for their antioxidant activity. Plant Foods and Human Nutrition 643945

24. J. B Harborne, 1973Phytochemical Methods. Chapman and Hall Ltd., London 49188

25. B. J Harborne, A. C Williams, Advances in flavonoids research since 1992Phytochemistry. 55481504

26. M Haupt, H Leithoff, D Meier, J Puls, H. D Richter, O Faix, 2003Heartwood extractives and natural durability of plantation-grown teakwood (Tectona grandis L.)- a case study. Holz als Roh- und Werkst. 616473474

27. L. F Hawley, L. C Fleck, C. A Richards, 1924The relation between durability and chemical composition in wood. Industrial & Engineering Chemistry. 167699700

28. B Hinterstoisser, B Stefke, M Schwanninger, 2000Wood: Raw material- material-Source of Energy for the future. Lignovisionen 22936

29. A Istek, H Sivrikaya, H Eroglu, S. K Gulsoy, 2005Biodegradation of Abies bornmülleriana (Mattf.) and Fagus orientalis (L.) by the white rot fungus Phanerochaete chrysosporium. International Biodeterioration & Biodegradation 556367

30. Kawaguchi, H, Kim, M, Ishida, M, Ahn, Y. J, Yamamoto, T, Yamaoka, R, Kozuka, M, Goto, K, & Takhashi, S. (1989). Several antifeedants fromPhellodendron amurense against Reticulitermes speratus. Agricultural Biology and Chemistry 5326352640

31. A. D Kokutse, A Stokes, H Bailleres, K Kokou, C Baudasse, 2006Decay resistance of Togolese teak (Tectona grandis L.) heartwood and relationship with colour. Trees 20, 219 223.

32. J Korb, 2007Termites. Current Biology 17995999

33. P. E Laks, P. A Mckaig, R. W Hemingway, 1988Flavanoid biocides: wood preservatives based on condensed tannins. Holzforschung 42299306

34. A. T Lelis, 1994Termite problem in São Paulo City-Brazil. In: Lenoir, A., Arnold, G., Lepage, M. (Eds.), Proceedings of the 12th Congress of the International Union for the Study of Social Insects-IUSSI, Paris4246

35. R. H. M. J Lemmens, 2008Dalbergia latifolia Roxb. In: Louppe, D.; Oteng-Amoako, A. A.; Brink, M. (Editors). Prota 7(1): Timbers/ Bois d'œuvre 1. [CD-Rom]. PROTA, Wageningen, Netherlands.

36. R. W Lotz, D. F Hollaway, 1988Wood preservation. US patent 4732817

37. C. A Maranhão, I. O Pinheiro, L. B. D. A Santana, L. S Oliveira, M. S Nascimento, L. W Bieber, 2013Antitermitic and antioxidant activities of heartwood extracts and main flavonoids of Hymenaea stigonocarpa Mart. International Biodeterioration & Biodegradation79913

38. Martínez-Sotres, C, López-Albarrán, P, Cruz-de-León, J, García-Moreno, T, Rutiagaquiñones, J. G, Vázquez-Marrufo, G, Tamariz-Mascarúa, J, & Herrera-Bucio, R. (2012). Medicarpin, an antifungal compound identified in hexane extract of Dalbergia congestiflora Pittier heartwood. International Biodeterioration & Biodegradation, 69, 38-40.

39. C. A Mcdaniel, 1992Major antitermitic components of the heartwood of Southern Catalpa. Journal of Chemical Ecology183359369

40. R Mitchell, T. D Sleeter, 1980Protecting wood from wood degrading organisms. US patent 4220688

41. M Morimoto, H Fukumoto, M Hiratani, W Chavasir, K Komai, 2006Insect Antifeedants, Pterocarpans and Pterocarpol in Heartwood ofPterocarpus macrocarpus Kruz. Biosci. Biotechnol. Biochem. 7018641868

42. Muñoz-mingarro, D, Acero, N, Llinares, F, Pozuelo, J. M, Galán de Mera, A, Vicenten, J. A, Morales, L, Alguacil, L. F, & Pérez, C. (2003). Biological activity of extracts from Catalpa bignonioides Walt. (Bignoniaceae) Journal of Ethnopharmacology, 87, 163-167.

43. T. C Scheffer, J. J Morell, 1998Natural Durability of Wood: A Worldwide Checklist of Species. Forest Research Laboratory, Oregon State University; College of Forestry, Research Contribution 2245

44. B Neya, M Hakkou, M Pétrissans, P Gérardin, 2004On the durability of Burkea Africana heartwood: evidence of biocidal and hydrophobic properties responsible for durability," Annals of Forest Science, 613277282

45. F. B Niamké, N Amusant, D Stien, G Chaix, Y Lozano, A. A Kadio, N Lemenager, D Goh, A. A Adima, S Kati-coulibaly, C Jay-allemand, 2012Dihydroxy-epiisocatalponol, a new naphthoquinone from Tectona grandis L.f. heartwood, and fungicidal activity. International Biodeterioration & Biodegradation 749398

46. W Ohmura, S Doi, M Aoyama, S Ohara, 2000Antifeedant activity of flavonoids and related compounds against the subterranean termiteCoptotermes formosanus Shiraki. Journal of Wood Science 46149153

47. Oliveira, L. S, Santana, A. L. B. D, Maranhão, C. A, Miranda, R. C. M, Galvão de Lima, V. L. A, Silva, S. I.; Nascimento, M. S, & Bieber, L. (2010). Natural resistance of five woods to Phanerochaete chrysosporium degradation. International Biodeterioration & Biodegradation 64711715

48. J. B Paes, V. M Morais, C. R Lima, 2005Resistência natural de nove madeiras do semi-árido brasileiro a fungos causadores da podridão-mole. R. Árvore293365371

49. R. C Pettersen, 1984The chemical composition of wood. In: Rowel, R.M. (Ed.), The Chemistry of Wood. Advances in Chemistry Series 207, American Chemical Society, Washington, DC, USA57126

50. S. P Pietarinen, S. M Willfor, F. A Virkstrom, B. R Holmbom, 2006Aspen knots, a rich source of flavonoids. Journal of Wood Chemistry and Technology 26245258

51. A Pizzi, M Merlim, 1981A new class of tannin adhesives for exterior particleboard. Int. J. Adhes. Adhes. 1, 261.

52. B Pizzo, C. L Pometti, J Charpentier, P Boizot, N Saidman, B. O. 2011Relationships involving several types of extractives of five native argentine wood species of genera Prosopis and Acacia. Industrial Crops and Products341851859

53. K. F Plomely, 1966Tannin-formaldehyde adhesives for wood. II Wattle tannin adhesives. CSIRO Division of Forest Products Technological Paper 39116

54. Ramos, I. E. C, Paes, J. B, Farias Sobrinho, D. W, & Santos, G. J. C. (2006). Efficiency of CCB on resistance of Prosopis juliflora (Sw.) D.C. wood in accelerated laboratory test decay. R. Árvore305811820

55. R Reyes-chilpa, N Viveros-rodriguez, F Gomez-garibay, D Alavez-solano, 1995Antitermitic activity of Lonchocarpus castilloi flavonoids and heartwood extracts. Journal of Chemical Ecology214455463

56. Reyes-chilpa, R, Gomez-Garibay, F, Moreno-Torres, G, Jimenez-Estrada, M, & Quiroz Vaásquez, R. I. (1998). Flavonoids and isoflavonoids with antifungal properties from Platymiscium yucatanum heartwood. Holzforschung525459462

57. M Royer, A. M. S Rodrigues, G Herbette, J Beauchêne, M Chevalier, B Hérault, B Thibaut, D Stiena, 2012Efficacy of Bagassa guianensis Aubl. extract against wood decay and human pathogenic fungi. International Biodeterioration & Biodegradation705559

58. R Sahai, S. K Agarwal, R. P Rastogi, 1980Auriculoside, a new flavan glycoside from acacia auriculiformis. Phytochemistry1915601562

59. Saka, S. (1991). Chemical composition and Distribution. Dekker, New York, 3-58.

60. W Sandermann, H. H Dietrichs, 1957Investigations on termite-resistant wood. Holz als Roh-und Werkstoff, 15, 281.

61. A. L. B. D Santana, C. A Maranhão, J. C Santos, F. M Cunha, G. M Conceição, L. W Bieber, M. S Nascimento, 2010Antitermitic activity of extractives from three Brazilian hardwoods against Nasutitermes corniger. International Biodeterioration & Biodegradation, 64, 7-12.

62. R. H Scheffrahn, 1991Allelochemical resistance of wood to termites. Sociobiology 19257281

63. T. C Scheffer, J. J Morrell, 1998Natural Durability of Wood: a Worldwide Checklist of Species. Oregon State University College of Forestry, Forest Research Laboratory Research Contribution 22.

64. T. P Schultz, W. B Harms, T. H Fisher, K. D Mcmurtrey, J Minn, D. D Nicholas, 1995Durability of angiosperm heartwood: The importance of extractives. Holzforschung4912934

65. T. P Schultz, D. D Nicholas, 2002Naturally durable heartwood: evidence for a proposed dual defensive function of the extractives. Phytochemistry 544752

66. Schultz, T. P, Hubbard Jr., T. F, Jin, L, Fisher, T. H, & Nicholas, D. D. (1990). Role of stilbenes in the natural durability of wood: fungicidal structure-activity relationships. Phytochemistry 2915011507

67. N Sekine, T Ashitani, T Murayama, S Shibutani, S Hattori, K Takahashi, 2009Bioactivity of latifolin and its derivatives against termites and fungi. Journal of Agricultural and Food Chemistry 5757075712

68. C. A Silva, M. B. B Monteiro, S Brazolin, G. A. C Lopez, A Richter, M. R Braga, 2007Biodeterioration of brazilwood Caesalpinia echinata Lam. (Leguminosae-Caesalpinioideae) by rot fungi and termites. International Biodeterioration & Biodegradation , 60, 285-292.

69. P Sirmah, S Dumarçay, P Gérardin, 2009Effect Unusual amount of (-)-mesquitol of from the heartwood of Prosopis juliflora Natural Product Research 23183189

70. P. K Sirmah, 2009Valorisation du «Prosopis juliflora» comme alternative à la diminution des ressources forestières au Kenya. Thesis-Université Henri Poincaré, Nancy I.

71. A Syofuna, A. Y Banana, G Nakabonge, 2012Efficiency of natural wood extractives as wood preservatives against termite attack. Maderas, Ciencia y Tecnología142155163

72. A. M Taylor, B. L Gartner, J. J Morrell, 2006Efects of Heartwood Extractive fractions of Thuja plicata and Chamaecyparis nootkatensison wood degradation by termites or Fungi. Journal of Wood Science52147153

73. M. F Thevenon, 1999Formulation of long-term, heavy-duty and lowtoxicwood preservatives. Application to the associations boric acid-condensed tannins and boric acid-proteins. Ph.D. Thesis, University of Nancy I, France.

74. Toirambe Bamoninga, B, & Ouattara, B. (2008). Morus mesozygia Stapf. In: Louppe, D, Oteng-Amoako, A. A, & Brink, M. (Editors). Prota 7(1): Timbers/Bois d'œuvre 1. [CD-Rom]. PROTA, Wageningen, Netherlands.

75. Toussaint, L. (1997). Utiliser les tannins pour la protection du bois. Telex bois, 4, 12

76. U Toshiaki, 2001Chemistry of extractives. In: «Wood and cellulosic chemistry». Ed Marcel Dekker, Inc. New York, 213241

77. Q. A Wang, B Zhou, Y Shan, 2004Progress on antioxidant activation and extracting technology of flavonoids. Chem. Product. Technol. 112933

78. Y Watanabe, R Mihara, T Mitsunaga, T Yoshimura, 2005Termite repellent sesquiterpenoids from Callitris glaucophylla heartwood. Journal of Wood Science 51514519

79. M Verma, S Sharma, R Prasad, 2010Biological alternatives for termite control: A review International Biodeterioration & Biodegradation 638959972

80. W. V Zucker, 1983Tannins: Does structure determine function? An ecological perspective. Am. Nat. 121335365

81. J. C. F Walker, 1993Primary Wood Processing. Principles and Practice.1st Edition. Chapman and Hall. 285 pp.

82. E Windeisen, G Wegener, G Lesnino, P Schumacher, 2002Investigation of the correlation between extractives content

and natural durability in 20 cultivated larch trees. Holz als Roh- und Werkstoff 60373374

83.   G. N Wolcott, 1955Organic termite repellents tested against Cryptotermes brevis. J. Agric. Univ. Puerto Rico, 39, 115.

Citations

CHAPTER 1

Dzeti Farhah Mohshim, Hilmi bin Mukhtar, Zakaria Man, and Rizwan Nasir, "Latest Development on Membrane Fabrication for Natural Gas Purification: A Review," Journal of Engineering, vol. 2013, Article ID 101746, 7 pages, 2013, doi:10.1155/2013/101746.

CHAPTER 2

Chaouki Ghenai, "Combustion of Syngas Fuel in Gas Turbine Can Combustor," Advances in Mechanical Engineering, vol. 2010, Article ID 342357, 13 pages, 2010. doi:10.1155/2010/342357.

CHAPTER 3

Jorge F. Gabitto and Costas Tsouris, "Physical Properties of Gas Hydrates: A Review," Journal of Thermodynamics, vol. 2010, Article ID 271291, 12 pages, 2010. doi:10.1155/2010/271291.

CHAPTER 4

Zuan Chen, Wuming Bai, Wenyue Xu, and Zhihe Jin, "An Analysis on Stability and Deposition Zones of Natural Gas Hydrate in Dongsha Region, North of South China Sea," Journal of Thermodynamics, vol. 2010, Article ID 185639, 6 pages, 2010, doi:10.1155/2010/185639.

CHAPTER 5

Juan Tomasini, Héctor de Santa Ana, Bruno Conti, et al., "Assessment of Marine Gas Hydrates and Associated Free Gas Distribution Offshore Uruguay," Journal of Geological Research, vol. 2011, Article ID 326250, 7 pages, 2011, doi:10.1155/2011/326250.

CHAPTER 6

R. Sivabalakrishnan and C. Jegadheesan, "Study of Knocking Effect in Compression Ignition Engine with Hydrogen as a Secondary Fuel," Chinese Journal of Engineering, vol. 2014, Article ID 102390, 8 pages, 2014. doi:10.1155/2014/102390.

CHAPTER 7

Elena Torralba-Calleja, James Skinner, and David Gutiérrez-Tauste, "CO2 Capture in Ionic Liquids: A Review of Solubilities and Experimental Methods," Journal of Chemistry, vol. 2013, Article ID 473584, 16 pages, 2013. doi:10.1155/2013/473584.

CHAPTER 8

Yun-Hoo Lee, Bong-Hwan Koh, Heung Soo Kim, and Myung Ho Song, "Compressive Strength Properties of Natural Gas Hydrate Pellet by Continuous Extrusion from a Twin-Roll System," Advances in Materials Science and Engineering, vol. 2013, Article ID 207867, 6 pages, 2013. doi:10.1155/2013/207867.

CHAPTER 9

M. S. Nascimento, A. L. B. D. Santana, C. A. Maranhão, L. S. Oliveira and L. Bieber (2013). Phenolic Extractives and Natural Resistance of Wood, Biodegradation - Life of Science, Dr. Rolando Chamy (Ed.), ISBN: 978-953-51-1154-2, InTech, DOI: 10.5772/56358.

Index